Electric Power Grid Reliability Evaluation

Electric Power Grid Reliability Evaluation

Models and Methods

Chanan Singh
Texas A&M University

Panida Jirutitijaroen
Press Quality Company

Joydeep Mitra
Michigan State University

IEEE PRESS

WILEY

Published by John Wiley & Sons, Inc., Hoboken, New Jersey.
Published simultaneously in Canada.

For general information on our other products and services or for technical support, please contact our Customer Care Department within the United States at (800) 762-2974, outside the United States at (317) 572-3993 or fax (317) 572-4002.

Wiley also publishes its books in a variety of electronic formats. Some content that appears in print may not be available in electronic formats. For more information about Wiley products, visit our web site at www.wiley.com.

Library of Congress Cataloging-in-Publication Data is available.

ISBN: 9781119486275

Cover design: Wiley
Cover image: © Ivan Kurmyshov/Shutterstock

Printed in the United States of America.

V10006067_111518

To Our Parents

The late Sadhu Singh and the late Pritam Kaur
Pairat Jirutitijaroen and Karuna Jirutitijaroen
Ajoy Kumar Mitra and Jharna Mitra

Contents

Preface

Over the past many years, the electric power grid has gone through transforma-tive changes. This has been driven by the need to reduce carbon emissions, have more monitoring to improve situational awareness and provide more choices and participative ability to the customers. The net result is that the grid is becoming more complex. Whereas the increased intelligence and capabilities built into the system provide opportunities for operating the grid in innova-tive ways that were not available earlier, they also introduce new possibilities of more problems resulting in possibilities of more widespread failures.

The power grid is an infrastructure that develops with time and involves decision-making that may be irreversible most of the time. For example, build-ing a transmission line or a wind or solar farm are not things that one can undo or change easily. So calculations to simulate the function of the new facilities and how they will affect the overall system has always been a part of the plan-ning and operation of power systems. It is for this reason that sophisticated analysis and simulation tools have been a part of these processes, and these tools have been going through transformations over time to suit new realities.

The same is true about the reliability analysis of power grid. The quantita-tive reliability evaluation makes it possible to do appropriate trade-offs with cost, emissions and other factors, resulting in a rational decision-making. The tools for power system reliability analysis have been evolving over time as more computational power has become available.

The material in this book has evolved through our teaching graduate and undergraduate classes to our students primarily at Texas A&M, Michigan State and National University of Singapore. The material has also been taught in short courses at industry and other academic institutions. The choice and pre-sentation of material is informed by our belief that a strong background in fundamentals is essential to understanding, properly adopting and improving the algorithms needed for reliability analysis. This is all the more important as the power system becomes more complex and its basic nature changes due to integration of renewable energy resources. More innovations in computational methods will be required as the need develops for adapting to new situations.

The material in this book is divided into two parts. The first part provides the theoretical foundations, covering a review of probability theory, stochastic processes and a frequency-based approach to understanding stochastic processes. These ideas are explained by using examples that connect with the power systems. Then both generic analytical and Monte Carlo methods are described. This first part can serve as material for a reliability course in general. The second part describes algorithms that have been developed for the reliability analysis of the power grid. This covers generation adequacy methods, and multinode analysis, which includes both multiarea as well as composite power system reliability evaluation. Then there are two chapters, one illustrating utilization of this material in energy planning and the second on integration of renewable resources that are characterized by their intermittent nature as energy sources.

Chanan Singh
Panida Jirutitijaroen
Joydeep Mitra

Acknowledgments

Many have contributed to this book in various ways, and we owe them a debt of gratitude.

Many of our students have contributed by attending our lectures, providing feedback and participating in the research that resulted in some of this material. It would be hard to thank them individually, so we thank them all as a group. We have also learned much by interacting with and following the work of our peers and other researchers and practitioners in the field, and we thank them all for enriching our expertise.

Last but not least, we would like to thank our families for their inspiration, support and patience while we were writing this book: our spouses Gurdeep, Wasu, and Padmini, our children Khenu Singh, Praow Jirutitijaroen, Ranadeep Mitra, Rukmini Mitra and Rajdeep Mitra, and grandchildren Kiran Parvan Singh and Saorise Vela Singh.

Chanan Singh
Panida Jirutitijaroen
Joydeep Mitra

Figures

Tables

PART I

Concepts and Methods in System Reliability

1

Introduction to Reliability

1.1 Introduction

The term reliability is generally used to relate to the ability of a system to perform its intended function. The term is also used in a more definite sense as one of the measures of reliability and indicates the probability of not failing by the end of a certain period of time, called the mission time. In this book, this term will be used in the former sense unless otherwise indicated. In a qualitative sense, planners and designers are always concerned with reliability, but the qualitative sense does not help us understand and make decisions while dealing with complex situations. However, when defined quantitatively it becomes a parameter that can be traded off with other parameters, such as cost and emissions.

There can be many reasons for quantifying reliability. In some situations, we want to know what the reliability level is in quantitative measures. For example, in military or space applications, we want to know what the reliability actually is, as we are risking lives. In commercial applications, reliability has a definite trade-off with cost. So we want to have a decision tool for which reliability needs to be quantified. The following example will illustrate this situation.

Example 1.1 A system has a total load of 500 MW. The following options are available for satisfying this load, which is assumed constant for simplicity:

5 generators, each with 100 MW;
6 generators, each with 100 MW;
12 generators, each with 50 MW.

The question we need to answer in terms of design and operation aspect is: *Which of these alternatives has the best reliability?*

A little thinking will show that there is no way to answer this question without some additional data on the stochastic behavior of these units, which are failure and repair characteristics. After we obtain this data, models can be built

Electric Power Grid Reliability Evaluation: Models and Methods, First Edition. Chanan Singh, Panida Jirutitijaroen, and Joydeep Mitra.
© 2019 by The Institute of Electrical and Electronic Engineers, Inc. Published 2019 by John Wiley & Sons, Inc.

to quantify the reliability for these three cases, and then the question can be answered.

1.2 Quantitative Reliability

Most of the applications of reliability modeling are in the steady state domain or in the sense of an average behavior over a long period of time. If we describe the system behavior at any instance of time by its state, the collection of possible states that the system may assume is called the *state space*, denoted by S.

In reliability analysis, one can classify the system state into two main categories, success or failure states. In success states the system is able to do its intended function, whereas in the failed states it cannot. We are mostly concerned with how the system behaves in failure states. The basic indexes used to characterize this domain are as follows.

Probability of failure
Probability of failure, denoted by p_f, is the steady state probability of the system being in the failed state or unacceptable states. It is also defined as the long run fraction of the time that system spends in the failed state. The probability of system failure is easily found by summing up the probability of failure states as shown in (1.1):

$$p_f = \sum_{i \in Y} p_i, \tag{1.1}$$

where

p_f system unavailability or probability of system failure;
Y set of failure states, $Y \subset S$;
S system state space.

Frequency of failure
Frequency of failure, denoted by f_f, is the expected number of failures per unit time, e.g., per year. This index is found from the expected number of times that the system transits from success states to failure states. As will be seen clearly in Chapter 4, this index can be easily obtained by finding the expected number of transitions across the boundary of subset Y of failure states.

Mean cycle time
Mean cycle time, denoted by T_f, is the average time that the system spends between successive failures and is given by (1.2). This index is simply the reciprocal of the frequency index:

$$T_f = \frac{1}{f_f}. \tag{1.2}$$

Mean down time

Mean down time, denoted by T_D, is the average time spent in the failed states during each system failure event. In other words, this is the expected time of stay in Y in one cycle of system up and down periods. This index can be found from (1.3):

$$T_D = \frac{p_f}{f_f}. \tag{1.3}$$

Mean up time

Mean up time, denoted by T_U, is the mean time that the system stays in the up states before system failure and is given by (1.4):

$$T_U = T_f - T_D. \tag{1.4}$$

There are several other indices that can be obtained as a function of the above indices, and these will be discussed in Chapter 5.

There are also applications in the time domain, say $[0, T]$. For example, at time 0, we may be interested in knowing the probability of not having sufficient generation at time T in helping decide the start of additional generation. The following indices could be used in such situations:

1. Probability of failure at time T
   This indicates the probability of being in the failed state at time T. This does not mean that the system did not fail before time T. The system may have failed before T and repaired, so this only indicates the probability of the system being in a failed state at time T.
2. Reliability for time T
   This is the probability that the system has not failed by time T.
3. Interval frequency over $[0, T]$
   This is the expected number of failures in the interval $[0, T]$.
4. Fractional duration
   This is the average probability of being in the failed state in interval $[0, T]$.

The most commonly computed reliability measures can be categorized as three indices as follows.

1. Expected value indexes: These indices involve
   Expected Power Not Supplied (EPNS) or Expected Unserved Energy (EUE).
2. Probability indices such as
   Loss of Load Probability (LOLP) or Loss of Load Expectation (LOLE).
3. Frequency and duration indices such as
   Loss of Load Frequency (LOLF) or Loss of Load Duration (LOLD).

1.3 Basic Approaches for Considering Reliability in Decision-Making

Having quantified the attributes of reliability, the next step is to see how it can be included in the decision process. There are perhaps many ways of doing it, but the most commonly used are described in this section. It is important to remember that the purpose of reliability modeling and analysis is not always to achieve higher reliability but to attain the required or optimal reliability.

Reliability as a constraint
Reliability can be considered a constraint within which other parameters can be changed or optimized. Until now this is perhaps the most common manner in which reliability considerations are implemented. For example, in generation reliability there is a widely accepted criterion of loss of load of one day in 10 years.

Reliability as a component of overall cost optimization
The conceptual relationship between cost and reliability can be appreciated from Figure 1.1. The overall cost is a combination of the investment cost and the cost of failures to the customers. The investment cost would tend to increase if we are interested in higher levels of reliability. The cost of failures to the customers, on the other hand, tends to decrease with increased level of reliability. If we combine these costs, the total cost is shown by the solid curve, which has a minimum value. The reliability at this minimum cost may be considered an optimal level; points to the left of this would be dominated by customer dissatisfaction, while points to the right may be dominated by investment cost considerations.

It can be appreciated that in this type of analysis we need to calculate the worth of reliability. In other words, how much do the customers think that

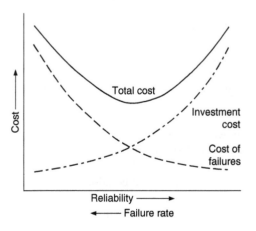

Figure 1.1 Trade-off between reliability and cost.

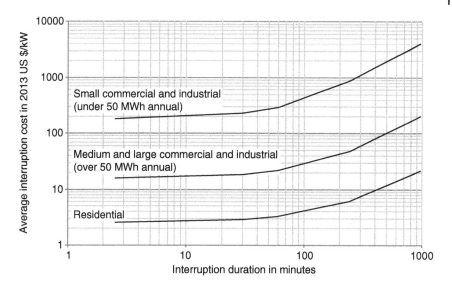

Figure 1.2 Customer damage function (compiled from data in [1]).

interruptions of power cost them? One way of doing this is through customer damage function, like the one shown in Figure 1.2.

The customer damage function provides the relationship between the duration of outage and the interruption cost in $/kW. The damage function is different depending on the type of customer. The damage function is clearly nonlinear with respect to the duration, increasing at much higher rates for longer outages. The frequency and duration indices defined earlier can be combined to yield the cost of interruptions using (1.5):

$$IC = \sum_{i=1}^{n} L_i f_i c_i(d_i),$$ (1.5)

where

n number of load points in the system;

L_i load requirement at load point i in kW;

f_i failure frequency at load point i in number of occurrence per year;

$c_i(d_i)$ customer damage function at load point i in $ per kW in terms of outage duration d_i;

d_i outage duration at load point i in hours.

Multi objective optimization and pareto-optimality

Generally there are conflicting objectives to be satisfied or optimized. For example, cost and reliability are conflicting objectives. Multi objective

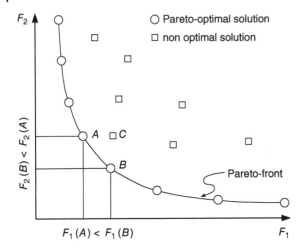

Figure 1.3 Multi objective optimization.

optimization, also known as multi criteria or multi attribute optimization, is the process of simultaneously optimizing two or more conflicting objectives subject to certain constraints. In multi objective optimization, Pareto-optimal solutions are usually derived, where the improvement of an objective will inevitably deteriorate at least another one. An example can be seen in Figure 1.3—given that lower values are preferred to higher values, point C is not on the Pareto frontier because it is dominated by both point A and point B; and points A and B are non inferior.

1.4 Objective and Scope of This Book

In general, reliability needs to be built, as far as possible, at the design or planning stage of a product or a system. Corrective actions to fix the reliability are generally more inconvenient and expensive. So far as the power grid is concerned, it is emerging as a highly complex system with heavy penetration of renewable energy sources, central and distributed energy storage and massive deployment of distributed communication and computational technologies allowing smarter utilization of resources. In addition, as the shape of the grid unfolds, there will be higher uncertainty in the planning and operation of these systems. As the complexity and uncertainty increase, the potential for possible failures with a significant effect on industrial complexes and society can increase drastically. In these circumstances maintaining the grid reliability and economy will be a very important objective and will be a challenge for those involved. Although many activities are involved in meeting these goals, educating the engineers in the discipline of reliability provides them with tools of analysis, trade-off and mental models for thinking. Reliability cannot be left to the goodwill of those designing or planning systems nor as a byproduct of these processes but must be engineered into the grid and its subsystems in a systematic and deliberate manner. An important step in this process is to model, analyze and predict the effect of design, planning and operating decisions on the reliability of the system. So there is a need for educational tools covering the spectrum of reliability modeling and evaluation tools needed for this emerging complex cyber-physical system.

The objective of this book is to provide state of the art tools for modeling and analyzing this system's reliability. This material will be useful for those who need to use these tools as well as those who want to do further research. They will be able to use this knowledge to make trade-offs between reliability, cost, environmental issues and other factors as needed.

To achieve this objective, we provide a strong background in general reliability that cultivates a deep understanding that can be used to develop appropriate tools as needed. We then use this foundation to build the tools for analyzing the power systems. The book can thus be used both by those who want to understand the tools of reliability analysis and those who want expertise in power system reliability.

1.5 Organization of This Book

We divide this book into two parts to meet its objective. The first part focuses on the basics of probability and stochastic processes as well as methods of reliability analysis developed based on these concepts. This part can be used by those who may be interested in learning about reliability methods in general. The second part develops models and methods that apply specifically to power system reliability.

2

Review of Probability Theory

2.1 Introduction

Knowledge of probability concepts is essential for power systems reliability modeling and analysis. These serve as fundamental ideas for the understanding of random phenomenon in reliability engineering problems. Probability theory is used to describe or model random occurrences in systems that behave according to probabilistic laws. Basic probability theory is reviewed in this chapter, with emphasis on application to power systems.

2.2 State Space and Event

Sample space or state space, usually denoted by S, is a collection of all possible outcomes of a random phenomenon. Consider the following examples:

⬦ Outcome of tossing a coin once: $S = \{\text{Head, Tail}\}$.
⬦ Outcome of rolling a dice: $S = \{1, 2, 3, 4, 5, 6\}$.
⬦ Status of a generator: $S = \{\text{Up, Down}\}$.
⬦ Status of two transmission lines: $S = \{(1U, 2U), (1U, 2D), (1D, 2U), (1D, 2D)\}$, where U denotes that a transmission line is working (i.e., in the up state) and D denotes that the transmission line has failed (and is in the down state), as shown in Figure 2.1.

In power system applications, we may want to focus our analysis on certain scenarios in the state space. For instance, in the example of the two transmission lines, we may be concerned only with the situation where at least one transmission line is working. This leads to what we call an *event*. An event is defined as a set of outcomes of a random phenomenon. It is a subset of a sample space. For example,

○ Rolling a dice yields a "1": $E = \{1\}$.
○ A generator has failed: $E = \{\text{Down}\}$.

Electric Power Grid Reliability Evaluation: Models and Methods, First Edition. Chanan Singh, Panida Jirutitijaroen, and Joydeep Mitra.

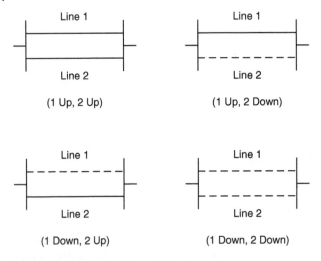

Figure 2.1 Status of two transmission lines.

- At least one transmission line is working, $E = \{(1U, 2U), (1U, 2D), (1D, 2U)\}$.
- Only one transmission line has failed, $E = \{(1U, 2D), (1D, 2U)\}$ as shown in Figure 2.2.

For any two events E_1 and E_2 in the state space S, the new event that contains outcomes from either E_1 or E_2 or both is called the *union of the events*, denoted

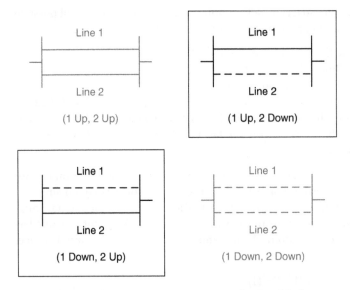

Figure 2.2 The event that only one transmission line has failed.

by $E_1 \cup E_2$. For example, if E_1 is an event that at least one transmission line is up, $E_1 = \{(1U, 2U), (1U, 2D), (1D, 2U)\}$, and E_2 is an event that at least one transmission line is down, $E_2 = \{(1U, 2D), (1D, 2U), (1D, 2D)\}$, then, union of event E_1 and E_2 is

$$E_1 \cup E_2 = \{(1U, 2U), (1U, 2D), (1D, 2U), (1D, 2D)\}.$$

For any two events E_1 and E_2 in the state space S, the new event that contains outcomes from both E_1 and E_2 is called the *intersection of the events*, denoted by $E_1 \cap E_2$. For example, the intersection of events E_1 and E_2 in the example of the two transmission lines is

$$E_1 \cap E_2 = \{(1U, 2D), (1D, 2U)\}.$$

There are cases where some events do not have any common outcome, i.e., the intersection of these events does not contain any outcome. Consider the event that both transmission lines are down, $E_3 = \{(1D, 2D)\}$; then the intersection of events E_1 and E_3 has no outcome. This null event is denoted by an empty set, $\emptyset$.

When the intersection of two events creates an empty set, the two events are said to be *mutually exclusive* or *disjoint events*. For example, if E_4 is an event that the two transmission lines are up, i.e., $E_4 = \{(1U, 2U)\}$, and E_5 is an event that two transmission lines are down, i.e., $E_5 = \{(1D, 2D)\}$, then it is impossible for E_4 and E_5 to happen together, and the intersection of E_4 and E_5 is a null set: $E_4 \cap E_5 = \emptyset$. We can conclude that E_4 and E_5 are mutually exclusive, and this is shown by the Venn diagram in Figure 2.3.

The concept of union and intersection of events can be extended to include more than two events. If $E_1, E_2, \ldots, E_n$ are events in the state space S, then union of these events, denoted by $\cup_{i=1}^{n} E_i$, is the event that contains outcomes from any of the events $E_1, E_2, \ldots, E_n$. The intersection of these events, which is defined in a similar way, denoted by $\cap_{i=1}^{n} E$, is an event that selects only outcome(s) that is (are) common in all the events $E_1, E_2, \ldots, E_n$. The same concept also applies when n goes to infinity.

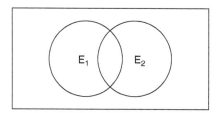 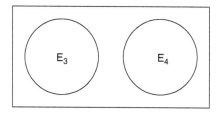

Figure 2.3 Venn diagram between inclusive events and mutually exclusive events.

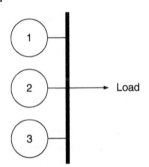

Figure 2.4 Three-generator system representation in Example 2.1.

Load

Example 2.1   Consider a system of three generators connected to a load, as shown in Figure 2.4. A generator can assume two statuses, either working in the up state or failure in the down state.

Let us find the possible outcomes (state space) of the status of generators in this problem, the event that the one generator is working and the event that satisfies any of the following criteria:

- One generator is working;
- Three generators failed;
- The third generator failed.

And lastly, find the event that satisfies all the above criteria. The state space of this problem is shown in Figure 2.5.

Let S be a state space of a status of three generating units. Then,

$$S = \{(1U, 2U, 3U), (1U, 2U, 3D), (1U, 2D, 3U), (1D, 2U, 3U),$$
$$(1D, 2U, 3D), (1D, 2D, 3U), (1U, 2D, 3D), (1D, 2D, 3D)\},$$

where U denotes a unit that is working and D denote a unit that is failed. The state space, S, shows the possible outcomes of this problem.

Let E_1 be an event that one generating unit is up; then

$$E_1 = \{(1D, 2U, 3D), (1D, 2D, 3U), (1U, 2D, 3D)\}.$$

Let E_2 be an event that three generating units are down then

$$E_2 = \{(1D, 2D, 3D)\},$$

and E_3 be an event that the third unit is down, then

$$E_3 = \{(1U, 2U, 3D), (1D, 2U, 3D), (1U, 2D, 3D), (1D, 2D, 3D)\}.$$

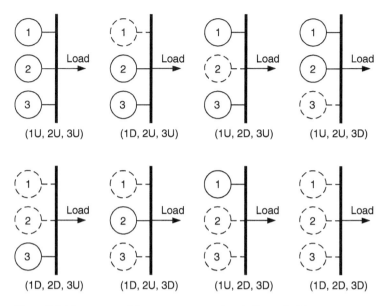

(1U, 2U, 3U)   (1D, 2U, 3U)   (1U, 2D, 3U)   (1U, 2U, 3D)

(1D, 2D, 3U)   (1D, 2U, 3D)   (1U, 2D, 3D)   (1D, 2D, 3D)

Figure 2.5 State space of three-generator system in Example 2.1.

The event that satisfies any one of the three criteria is given as the union of E_1, E_2, E_3 or

$$E_1 \cup E_2 \cup E_3 = \{(1U, 2U, 3D), (1D, 2U, 3D), (1D, 2D, 3U), (1U, 2D, 3D),$$
$$(1D, 2D, 3D)\}$$

The event that satisfies all of the three criteria is given as the intersection of E_1, E_2, E_3 or,

$$E_1 \cap E_2 \cap E_3 = \{(1D, 2D, 3D)\}.$$

Let us also find the event, denoted by E_3^c, that the third unit is up; then,

$$E_3^c = \{(1U, 2U, 3U), (1U, 2D, 3U), (1D, 2U, 3U), (1D, 2D, 3U)\}.$$

Note that this event E_3^c contains all possible outcomes in the state space that are not in the event E_3, which describes the outcomes that the third unit is down.

We can now define a new event, denoted by E^c, a *complement of an event E*, to be the set of outcomes that are in the state space, S, but not included in an event E. This means that E^c will occur only when E does not occur. This also implies that E and E^c are mutually exclusive ($E \cap E^c = \emptyset$) and that the union of E and E^c yields the state space, $E \cup E^c = S$.

2.3 Probability Measure and Related Rules

Probability is defined as a quantitative measure of an event E in a state space S. This measure is denoted by $P(E)$, called probability of an event E and defined to satisfy the following properties.

1. $0 \leq P(E) \leq 1$
2. $P(S) = 1$
3. If $E_1, E_2, \ldots$ are mutually exclusive events in S, then $P(\cup_{i=1}^{\infty} E_i) = \sum_{i=1}^{\infty} P(E_i)$.

For engineering applications, probability can be interpreted as a measure of how frequent an event will occur in a long-run experiment. Intuitively, this measure of an event should be proportional to a number of times that an outcome in the event occurs divided by the total number of experiments. For example,

- If a coin is tossed once, what is the probability of the outcome being a Head? If the coin is fair, i.e., there is equal chance to appear as Head or Tail,

  $$P(\{Head\}) = P(\{Tail\}).$$

  Since $S = \{Head, Tail\}$ and $P(S) = 1$, the probability of being Head is

  $$P(\{Head\}) = \frac{1}{2}.$$

- If a dice is rolled once, what is the probability of the outcome being "1"? If the dice is fair, then each number has the same chance to appear,

  $$P(\{1\}) = P(\{2\}) = P(\{3\}) = P(\{4\}) = P(\{5\}) = P(\{6\}).$$

  Using the 2nd and 3rd properties, we have

  $$P(\{i\}) = \frac{1}{6}, i \in \{1, 2, \ldots, 6\}.$$

  The probability of rolling "1" is $P(\{1\}) = \frac{1}{6}$.

- If a dice is rolled once, what is the probability of the outcome being an odd number?
  The event of being odd number is $\{1, 3, 5\}$. Since these events are mutually exclusive, using the 3rd property, we have

  $$P(\{1, 3, 5\}) = P(\{1\}) + P(\{2\}) + P(\{3\}) = \frac{1}{2}.$$

For any set, E, the union of E and its complement, E^c, yields the state space, $E \cup E^c = S$. Since E and E^c are always mutually exclusive, by properties 2 and 3, we have $P(E \cup E^c) = P(E) + P(E^c) = P(S) = 1$. This implies that $P(E^c) = 1 - P(E)$. For some application problems, it is easier to calculate probability of a complement of an event than the probability of an event itself. We can use this property, called *complementation rule*, to help us find a probability of an event.

We can also derive another important rule, called *addition rule*, to find a probability of union of two events. If the two events are mutually exclusive, we arrive at the same result as the 3rd property. We now consider the case when the two events are not disjoint and introduce the concept by the following example.

Example 2.2 Consider a system of two transmission lines. If E_1 is an event that at least one transmission line is up, $E_1 = \{(1U, 2U), (1U, 2D), (1D, 2U)\}$, E_2 is an event that at least one transmission line is down, $E_2 = \{(1U, 2D), (1D, 2U), (1D, 2D)\}$; then, union of event E_1 and E_2 is

$$E_1 \cup E_2 = \{(1U, 2U), (1U, 2D), (1D, 2U), (1D, 2D)\},$$

and the intersection of event E_1 and E_2 is

$$E_1 \cap E_2 = \{(1U, 2D), (1D, 2U)\}.$$

If we add probability of E_1 to probability of E_2, we have

$$P(E_1) + P(E_2) = P(\{(1U, 2U), \underline{(1U, 2D), (1D, 2U)}, (1U, 2D), (1D, 2U), (1D, 2D)\}).$$

Note that the events $\{(1U, 2D), (1D, 2U)\}$ appear twice. Rearranging the events, we have

$$P(E_1) + P(E_2) = P(\{(1U, 2U), (1U, 2D), (1D, 2U), (1D, 2D)\})$$
$$+ P(\{(1D, 2U), (1U, 2D)\}).$$

This means that

$$P(E_1) + P(E_2) = P(E_1 \cup E_2) + P(E_1 \cap E_2).$$

For any two events, we can calculate the probability of union of two events as follows.

$$P(E_1 \cup E_2) = P(E_1) + P(E_2) - P(E_1 \cap E_2). \tag{2.1}$$

Note that if E_1 and E_2 are mutually exclusive, then $P(E_1 \cup E_2) = P(E_1) + P(E_2)$ follows from the 3rd condition. This implies $P(E_1 \cap E_2) = 0$, and we can conclude that $P(\emptyset) = 0$ since $E_1 \cap E_2 = \emptyset$.

In general, the probability of union of n events can be found as shown below:

$$P(E_1 \cup E_2 \cup \ldots \cup E_n) = \sum_i P(E_i) - \sum_{i<j} P(E_i \cap E_j) \tag{2.2}$$
$$+ \sum_{i<j<k} P(E_i \cap E_j \cap E_k) - \ldots$$
$$+ (-1)^{n-1} P(E_1 \cap E_2 \cap \ldots \cap E_n).$$

Table 2.1 Probability of an Event in Example 2.3

Event	Probability
$(1U, 2U)$	0.81
$(1U, 2D)$	0.09
$(1D, 2U)$	0.09
$(1D, 2D)$	0.01

Consider again a system of two transmission lines. Suppose that the first transmission line fails, and we may wish to know the probability of the second line failing. Let E_1 be an event that the first transmission line fails, $E_1 = \{(1D, 2U), (1D, 2D)\}$, and $E_2|E_1$ be an event that the second transmission line fails, given that one transmission line has already failed. The event that the second transmission line fails is $E_2 = \{(1U, 2D), (1D, 2D)\}$. However, $(1U, 2D)$ cannot occur in this problem since the first transmission line has already failed. We are interested in calculating the *conditional probability* that the event E_2 occurs, given that the event E_1 has already occurred. We denote this probability as $P(E_2|E_1)$.

Intuitively, when E_1 has already occurred, we can only consider the states with occurrences E_1 in the state space. This means that the state space has shrunk to become the set E_1, and the events in E_2 will have to be in common with the events in E_1. This leads to the following formula for conditional probability:

$$P(E_2|E_1) = \frac{P(E_2 \cap E_1)}{P(E_1)}. \tag{2.3}$$

This formula is properly defined only when $P(E_1) > 0$.

Example 2.3 Consider the same system as in Example 2.2. The state space of this problem is $S = \{(1U, 2U), (1U, 2D), (1D, 2U), (1D, 2D)\}$. The probability of each event is given as shown in Table 2.1. Let us calculate the probability that the second transmission line fails given that the first transmission line has already failed.

Let E_1 be an event that the first transmission line fails, $E_1 = \{(1D, 2U), (1D, 2D)\}$, and E_2 be an event that the second transmission line fails, $E_2 = \{(1U, 2D), (1D, 2D)\}$. We need to calculate $P(E_2|E_1)$. Figure 2.6 shows the state space of this example.

Since $E_2 \cap E_1 = \{(1D, 2D)\}$, $P(E_2 \cap E_1) = 0.01$ and $P(E_1) = 0.09 + 0.01 = 0.1$, then we have

$$P(E_2|E_1) = \frac{P(E_2 \cap E_1)}{P(E_1)} = \frac{0.01}{0.1} = 0.1.$$

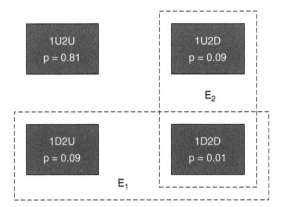

Figure 2.6 State space of two-transmission line in Example 2.3.

This conditional probability rule is in fact a very powerful technique to help us calculate the probability of an event in a state space. To illustrate how we can apply this technique, let us first divide the state space into two mutually exclusive sets, $S = B_1 \cup B_2$, $B_1 \cap B_2 = \emptyset$. An event E in the state space has to be inclusive with either event B_1 or B_2 and can be described as $E = (E \cap B_1) \cup (E \cap B_2)$. Using addition rule,

$$P(E) = P(E \cap B_1) + P(E \cap B_2) - P(E \cap E \cap B_1 \cap B_2).$$

Since $B_1 \cap B_2 = \emptyset$, it follows that $P(E \cap E \cap B_1 \cap B_2) = 0$. From conditional probability rule,

$$P(E \cap B_1) = P(E|B_1) \times P(B_1),$$
$$P(E \cap B_2) = P(E|B_2) \times P(B_2).$$

Then,

$$P(E) = P(E|B_1) \times P(B_1) + P(E|B_2) \times P(B_2).$$

This expression can be interpreted as a weighted average of the conditional probability of E on a given event when the weight is the probability of the event in which E is conditioned to occur.

For n mutually exclusive events, B_i that $\cup_{i=1}^{n} B_i = S$. We can find a probability of any event E from the following *Bayes' rule.*

$$P(E) = \sum_{i=1}^{n} P(E|B_i) \times P(B_i) \tag{2.4}$$

For example, if the state space is divided into five mutually exclusive events, $B_1, B_2, \dots, B_5$, then the probability of an event E can be found from conditional

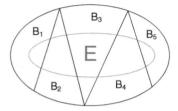

Figure 2.7 Graphical representation of Bayes' rule.

probability of event E on each of the disjoint events and can be shown in graphical form in Figure 2.7.

The conditional probability rule can also be used to calculate the probability of an intersection of events. Recall that for any two events,

$$P(E_2 \cap E_1) = P(E_2|E_1) \times P(E_1).$$

We now define another important property of two events called *independent*. The two events are independent, if and only if,

$$P(E_2 \cap E_1) = P(E_2) \times P(E_1). \tag{2.5}$$

This also implies that $P(E_2|E_1) = P(E_2)$, which means that the probability that E_2 will occur does not depend on whether E_1 has already occurred or not. This property helps us to calculate the probability of the intersection of two events by simply multiplying the probabilities of the two events. We call this *multiplication rule*. It should be noted that the independence property is different from mutually exclusive property of two events and cannot be described using a Venn diagram.

Example 2.4   Consider the same system as in Example 2.3. Determine whether or not the event of failure of the first transmission line and the event of failure of the second transmission line are independent. Let E_1 and E_2 be events that the first transmission line and the second transmission line fail, then

$$E_1 = \{(1D, 2U), (1D, 2D)\},$$
$$E_2 = \{(1U, 2D), (1D, 2D)\}.$$

We can find $P(E_1) = 0.09 + 0.01 = 0.1$, and $P(E_2) = 0.09 + 0.01 = 0.1$.

Since $E_2 \cap E_1 = \{(1D, 2D)\}$, $P(E_2 \cap E_1) = 0.01$. Then, we have $P(E_2 \cap E_1) = P(E_2) \times P(E_1) = 0.01$.

This means that the event that the first transmission line fails and the event that the second transmission line fails in this problem are independent.

Now, if the probability of this state space has changed and is given by the Table 2.2, determine if the two events are still independent or not.

Table 2.2 Probability of an Event in Example 2.4

Event	Probability
$(1U, 2U)$	0.80
$(1U, 2D)$	0.10
$(1D, 2U)$	0.09
$(1D, 2D)$	0.01

In this case, $P(E_1) = 0.09 + 0.01 = 0.1$, $P(E_2) = 0.10 + 0.01 = 0.11$ and $P(E_2 \cap E_1) = 0.01$. Then,

$$P(E_2 \cap E_1) = 0.01 \neq P(E_2) \times P(E_1) = 0.011.$$

This shows that the two events are not independent. This implies that the event that the first transmission line fails is dependent on the event that the second transmission line fails and vice versa.

From Example 2.4, since a transmission line can assume either up or down state, i.e., the state space of one transmission line is $S = \{U, D\}$. This means that if the failure probability of a transmission line is $P(D) = 0.1$, the probability that a transmission line will work is $P(U) = 1 - P(D) = 0.9$ according to the complementary rule.

For a system of two identical transmission lines, let E be an event that the first transmission line is working and the second transmission line fails, $E = \{(1U, 2D)\}$. If the working or failure statuses of the two transmission lines are independent, then the probability of this event can be found from multiplication rule,

$$P(E) = P(1U \cap 2D) = P(1U) \times P(2D) = 0.09,$$

which is the same as shown in Example 2.3.

Example 2.5 Consider the same system of three generators connected to a load as shown in Example 2.1. Assume that each generator has 50 MW capacity with the probability of failure of 0.01, and each generator fails independently. Let us find the probability that the system will supply 0, 50, 100 and 150 MW to the load and the probability of loss of load when the load is 50, 100 or 150 MW with 0.20, 0.75 and 0.05 probability accordingly.

Let us first define the events as follows:

E_1 Event that the system will supply 0 MW.
E_2 Event that the system will supply 50 MW.

E_3 Event that the system will supply 100 MW.
E_4 Event that the system will supply 150 MW.

These events are given in the following:

$$E_1 = \{(1D, 2D, 3D)\}$$
$$E_2 = \{(1U, 2D, 3D), (1D, 2U, 3D), (1D, 2D, 3U)\}$$
$$E_3 = \{(1U, 2U, 3D), (1D, 2U, 3U), (1U, 2D, 3U)\}$$
$$E_4 = \{(1U, 2U, 3U)\}.$$

In order to calculate the probability of each event, it is important to note that each generator has failure probability of 0.01. Since all generators fail independently, we can use multiplication rule:

$$P(E_1) = P(1D \cap 2D \cap 3D) = P(1D) \times P(2D) \times P(3D) = 0.00001$$
$$\begin{aligned}P(E_2) &= P\{(1U, 2D, 3D) \cup (1D, 2U, 3D) \cup (1D, 2D, 3U)\}\\ &= P(1U \cap 2D \cap 3D) + P(1D \cap 2U \cap 3D) + P(1D \cap 2D \cap 3U)\\ &= \{P(1U) \times P(2D) \times P(3D)\} + \{P(1D) \times P(2U) \times P(3D)\}\\ &\quad + \{P(1D) \times P(2D) \times P(3U)\}\\ &= 0.000297\end{aligned}$$
$$\begin{aligned}P(E_3) &= P\{(1U, 2U, 3D) \cup (1D, 2U, 3U) \cup (1U, 2D, 3U)\}\\ &= P(1U \cap 2U \cap 3D) + P(1D \cap 2U \cap 3U) + P(1U \cap 2D \cap 3U)\\ &= \{P(1U) \times P(2U) \times P(3D)\} + \{P(1D) \times P(2U) \times P(3U)\}\\ &\quad + \{P(1U) \times P(2D) \times P(3U)\}\\ &= 0.029403\end{aligned}$$
$$\begin{aligned}P(E_4) &= P(1U \cap 2U \cap 3U)\\ &= P(1U) \times P(2U) \times P(3U)\\ &= 0.970299.\end{aligned}$$

The loss of load can occur in three mutually exclusive load scenarios, i.e., when the load is 50, 100 or 150 MW. First, we define the following events.

F Event of loss of load.
B_1 Event that load is 50 MW.
B_2 Event that load is 100 MW.
B_3 Event that load is 150 MW.

Then, the probability of loss of load can be found using Bayes' rule as follows.

$$P(F) = P(F|B_1) \times P(B_1) + P(F|B_2) \times P(B_2) + P(F|B_3) \times P(B_3).$$

The loss of load event given that the load is 50 MW, described by Figure 2.8, will occur when all units fail. Thus, $P(F|B_1) = P(1D \cap 2D \cap 3D) = P(E_1) = 0.00001$.

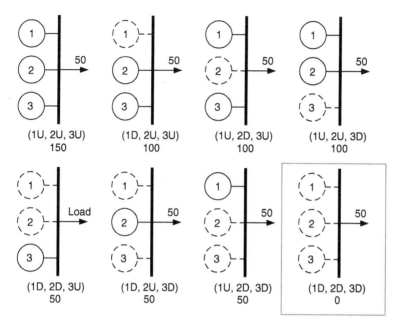

Figure 2.8 State space representation for loss of load event when load is 50 MW.

The loss of load event given that the load is 100 MW will occur when at least two units fail. This event is described by Figure 2.9. Thus, $P(F|B_2) = P\{(1U, 2D, 3D) \cup (1D, 2U, 3D) \cup (1D, 2D, 3U) \cup (1D, 2D, 3D)\} = P(E_2) + P(E_1) = 0.000298$.

The loss of load event given that the load is 150 MW will occur when at least one unit fails. This event is described by Figure 2.10. Equivalently, no loss of load will occur when all units are working. Using complementary rule, $P(F|B_3) = 1 - P(1U, 2U, 3U) = 1 - P(E_4) = 0.029701$.

From Bayes' rule, we can calculate the loss of load probability as follows.

$$P(F) = P(F|B_1) \times 0.20 + P(F|B_2) \times 0.75 + P(F|B_3) \times 0.05 = 0.00170875.$$

We can also use Bayes' rule to calculate loss of load probability by conditioning on the delivered capacity of the three generators instead of conditioning on the different load levels shown in this example.

As seen in Example 2.5, we are interested in knowing the total generating capacity of the system rather than the status of each generator. This real-valued quantity is more important for our analysis since we are interested in a *function of the outcome* (generating capacity) of the event rather than the *outcome* (status

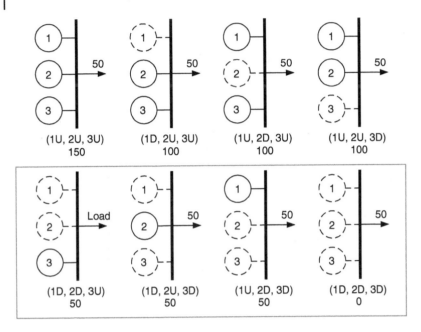

Figure 2.9 State space representation for loss of load event when load is 100 MW.

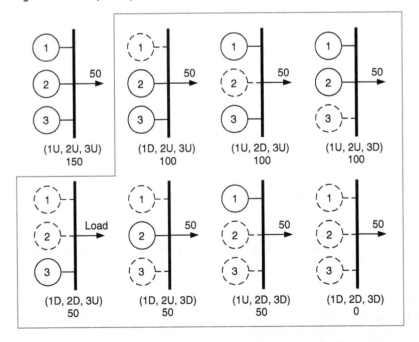

Figure 2.10 State space representation for loss of load event when load is 150 MW.

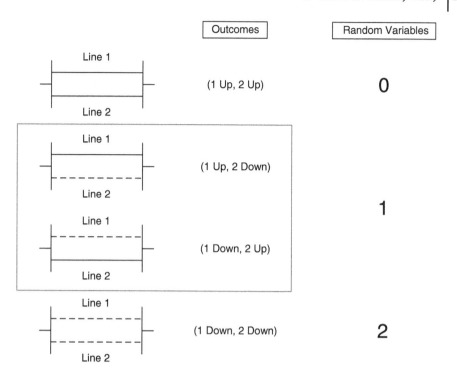

Figure 2.11 Example of discrete random variable.

of each generator) itself. This quantity is a real-valued function defined on the sample space and is called a *random variable*.

2.4 Random Variables

A random variable is a real-valued function that assigns numerical values to all outcomes in the state space. A random variable can take on either countable real values or continuous real values. We call the random variables with countable values discrete random variables, and those with continuous real values continuous random variables. For example:

- Discrete random variable: Number of failed transmission lines in the system with the state space of two transmission lines shown in Figure 2.11.
- Continuous random variable: Time to failure of a generator shown in Figure 2.12.

Since each outcome is associated with a probability measure, we can assign probabilities to any possible value of the random variable. For example,

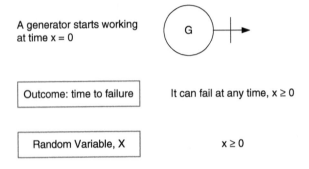

A generator starts working at time x = 0

G

Outcome: time to failure   It can fail at any time, x ≥ 0

Random Variable, X   x ≥ 0

Figure 2.12 Example of continuous random variable.

consider the system of two transmission lines if the random variable is the number of failed transmission lines and the probability of the outcomes is the same as given in Example 2.3. Then

- $P\{X = 0\} = P(\{(1U, 2U)\}) = 0.81$,
- $P\{X = 1\} = P(\{(1U, 2D), (1D, 2U)\}) = 0.9 + 0.9 = 0.18$,
- $P\{X = 2\} = P(\{(1D, 2D)\}) = 0.01$.

Note that $P\{X = 0\} + P\{X = 1\} + P\{X = 2\} = 1$.

2.4.1 Probability Density Function

A discrete random variable assumes only countable values of $x_1, x_2, \ldots, x_n$ from the set of real numbers. The function that gives probabilities associated with all possible values of a discrete random variable, X, is called *probability mass function*, denoted by

$$p(x) = P\{X = x\}. \tag{2.6}$$

The following are the properties of probability mass function of a discrete random variable:

1. $0 \leq p(x_i) \leq 1, i = 1, 2, \ldots, n$
2. $p(x_i) = 0$ if $i \notin \{1, 2, \ldots, n\}$
3. $\sum_{i=1}^{\infty} p(x_i) = \sum_{i=1}^{\infty} P\{X = x_i\} = 1$.

A graphical representation of a probability mass function is shown in Figure 2.13. In this example, a discrete random variable can assume any of the values of 1, 2, 3, 4 or 5 with 0.1, 0.2, 0.3, 0.3 and 0.1 probabilities accordingly.

Similarly, for a continuous random variable X, we define a non negative function, $f(x)$, called a *probability density function*, for all real numbers

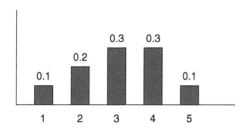

Figure 2.13 Example of probability mass function of a discrete random variable.

$x \in (-\infty, \infty)$. A function $f(x)$ has to satisfy the criteria that for any set A of real numbers $x \in A$,

$$\int_A f(x)dx = P\{X \in A\}. \tag{2.7}$$

A continuous random variable assigns real values to all outcomes in the state space; thus the probability of all real values has to add up to one, which is given as

$$\int_{-\infty}^{\infty} f(x)dx = P\{X \in (-\infty, \infty)\} = 1. \tag{2.8}$$

Example 2.6 Let X denote a continuous random variable representing time to failure (in days) of a generator by probability density function $f(x)$; what is the probability that a generator will fail during the 2nd and 3rd days, $A = [2, 3]$? This probability can be found from

$$P\{X \in [2, 3]\} = P\{2 \le X \le 3\} = \int_2^3 f(x)dx.$$

Note that with this definition, the probability that a continuous random variable will assume any particular value a will be zero since $P\{X = a\} = \int_a^a f(x)dx = 0$.

2.4.2 Probability Distribution Function

A random variable can also be characterized by a *cumulative distribution function* or *distribution function*, denoted by $F(a)$, which gives the probability that a random variable takes on value less than or equal to a real number a.

$$F(a) = P\{X \le a\} \tag{2.9}$$

For a discrete random variable,

$$F(a) = \sum_{x_i \le a} P\{X = x_i\} = \sum_{x_i \le a} p(x_i). \tag{2.10}$$

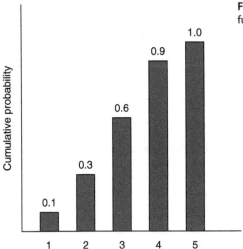

Figure 2.14 Example of distribution function of a discrete random variable.

A graphical representation of a distribution function is shown in Figure 2.14. In this example, a discrete random variable can assume values of 1, 2, 3, 4 or 5 with corresponding probabilities of 0.1, 0.2, 0.3, 0.3 and 0.1.

For a continuous random variable,

$$F(a) = P\{X \in (-\infty, a)\} = \int_{-\infty}^{a} f(x)dx. \tag{2.11}$$

From the above expression, a distribution function $F(a)$ is a non decreasing function of a, $\lim_{a\to\infty} F(a) = F(\infty) = P\{X \in (-\infty, \infty)\} = 1$ and $\lim_{a\to-\infty} F(a) = F(-\infty) = P\{X \in (-\infty, -\infty)\} = 0$.

Example 2.7 Let X denote a continuous random variable representing time to failure (in days) of a generator by a distribution function $F(x)$; what is the probability that a generator will fail during the 2nd and 3rd days, $A = [2, 3]$?

This probability can be found from

$$\begin{aligned} P\{X \in [2,3]\} &= P\{2 \le X \le 3\} \\ &= P\{X \in (-\infty, 3)\} - P\{X \in (-\infty, 2)\} \\ &= F(3) - F(2). \end{aligned}$$

From (2.11), if we differentiate both sides, we have

$$\frac{dF(a)}{da} = F'(a) = f(a). \tag{2.12}$$

This means that the probability density function can be found from differentiating the distribution function. Equivalently, we can also write

$$f(a) = \lim_{\Delta a \to 0} \frac{F(a + \Delta a) - F(a)}{\Delta a} = \lim_{\Delta a \to 0} \frac{P\{a \leq X \leq a + \Delta a\}}{\Delta a}. \qquad (2.13)$$

2.4.3 Survival Function

Consider a continuous random variable representing, for example, time to failure of a component. The probability density function $f(x)$ of this random variable can give the probability of failure at a certain time. The probability distribution function $F(a)$ gives the probability that it will fail within time a. In reliability analysis, it is sometimes more interesting to know the probability that the component will fail beyond a specified time. In this case it is more convenient to work with the complementary of the distribution function called *survival function*.

Survival function, denoted by $R(a)$, gives the probability of a component surviving beyond time a.

$$R(a) = P\{X > a\} \qquad (2.14)$$

when X is a random variable of time to failure of the component.

Survival function can be determined from either the probability density function or the probability distribution function, as shown below:

$$R(a) = P\{X > a\} = 1 - P\{X \leq a\} = 1 - F(a) \qquad (2.15)$$

$$R(a) = P\{X > a\} = \int_a^\infty f(x)dx. \qquad (2.16)$$

It follows from (2.12) that

$$\frac{d(1 - R(a))}{da} = -R'(a) = f(a). \qquad (2.17)$$

Any one of the density function, distribution function or survival function can be determined from the others. In other words, we can denote a random variable with one of the three functions interchangeably.

2.4.4 Hazard Rate Function

Another function used extensively in reliability analysis is called *hazard rate function*, denoted by $h(a)$. It is widely used to describe a random variable X representing time to failure of a component.

A hazard rate function is a function that gives a rate at time a at which a component fails given that it has survived for time a. This function is the rate of a conditional probability of failure at time a and is given by (2.18):

$$h(a) = \lim_{\Delta a \to 0} \frac{P\{a \le X \le a + \Delta a | X > a\}}{\Delta a} \tag{2.18}$$

As $\Delta a \to 0$, the hazard rate function can be written as $h(a)\Delta a = P\{a \le X \le a + \Delta a | X > a\}$. This gives a conditional probability of a random variable taking a value in the interval $[a, a + \Delta a]$ given that the value is greater than a. When X represents time to failure of a component, $h(a)\Delta a$ gives the probability that a component will fail during interval $[a, a + \Delta a]$ given that it has been working (not failed) up to time a.

Depending on the context of usage, a hazard rate function is known by a variety of names, such as age specific failure rate, failure rate, repair rate and force of mortality. We can find this function from the density and survival function as follows. Using conditional probability rule,

$$\begin{aligned} h(a) &= \lim_{\Delta a \to 0} \frac{P\{a \le X \le a + \Delta a \cap X > a\}}{\Delta a} \times \frac{1}{P\{X > a\}} \\ &= \lim_{\Delta a \to 0} \frac{P\{a \le X \le a + \Delta a\}}{\Delta a} \times \frac{1}{P\{X > a\}} \\ &= \frac{f(a)}{R(a)}. \end{aligned} \tag{2.19}$$

Using (2.17), we have

$$h(a) = \frac{-R'(a)}{R(a)} = -\frac{d}{da}[\ln R(a)]. \tag{2.20}$$

Integrating (2.20) yields

$$R(a) = e^{-\int_0^a h(x)dx}. \tag{2.21}$$

In addition,

$$f(a) = h(a)e^{-\int_0^a h(x)dx}. \tag{2.22}$$

Equations (2.21) and (2.22) allow us to uniquely determine the probability density function and survival function from the hazard rate function. All three functions can be used to calculate one another. Their mathematical relationship can be derived and is shown in Figure 2.15.

Probability density function

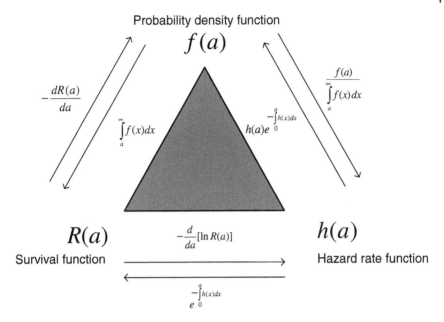

Figure 2.15 Triangle defining relationship between density function, survival function and hazard rate function.

2.5 Jointly Distributed Random Variables

The previous sections only consider a single random variable. Sometimes we need to consider the joint probabilities behavior of two or more random variables. As an example, consider a power system consisting of generators and transmission lines. In this case, both generating capacity and transmission line capabilities exhibit probabilistic behavior. If we need to find out the total available capacity of the system, we therefore need to describe the two uncertainties using two random variables with some distribution functions. This leads us to consider the situation of two or more random variables.

Consider two random variables X and Y. We define a *joint probability distribution function* of these two random variables as follows:

$$F(a, b) = P\{X \le a, Y \le b\}, \tag{2.23}$$

where $a, b \in (-\infty, \infty)$.

Similar to the single random variable cases, we define the joint probability density function as follows. For discrete case random variables X and Y, the joint probability density function is given below:

$$p(x, y) = P\{X = x, Y = y\}. \tag{2.24}$$

For continuous random variables X and Y, we define a non negative joint probability density function, $f(x, y)$, for all real numbers $x, y \in (-\infty, \infty)$. A function $f(x, y)$ has to satisfy the criteria that for any set A, B of real number $x \in A, y \in B$,

$$\int_B \int_A f(x, y) dx dy = P\{X \in A, Y \in B\}. \tag{2.25}$$

If we know the joint probability density function of X and Y, we can find a probability density function of X since $y \in (-\infty, \infty)$, we have

$$f(x) = \int_{-\infty}^{\infty} f(x, y) dy. \tag{2.26}$$

Similarly, a probability density function of Y is

$$f(y) = \int_{-\infty}^{\infty} f(x, y) dx. \tag{2.27}$$

If the two random variables are independent, then from (2.5) we can write

$$F(a, b) = P\{X \leq a, Y \leq b\} = P\{X \leq a \cap Y \leq b\} = P\{X \leq a\}P\{Y \leq b\},$$

which also implies $f(x, y) = f(x)f(y)$.

2.6 Expectation, Variance, Covariance and Correlation

A random variable can be expressed by density function, survival function or hazard rate function. These functions yield the probability associated with a real valued variable representing different outcomes in the state space. It is sometimes of interest to represent a random variable X by a single value. This value is called *expectation* or *expected value*, denoted by $E[X]$ or μ.

An expected value is an average of real possible values that a random variable assumes randomly for a long run experiment. For a discrete random variable X, we have

$$E[X] = \sum_i x_i P\{X = x_i\}. \tag{2.28}$$

This expectation is therefore the weighted sum of all discrete real values x_i, each x_i weighted by the probability of X assuming the value x_i.

For a continuous random variable X having density function $f(x)$, we have

$$E[X] = \int_{-\infty}^{\infty} x f(x) dx. \tag{2.29}$$

It can be verified from both (2.28) and (2.29) that for a random variable X, $E[aX + b] = aE[X] + b$. The expected value of a summation of random variable can be found as follows:

$$E\left[\sum_i X_i\right] = \sum_i E[X_i].$$ (2.30)

Suppose that we want to calculate the expected value of a function $g(.)$ of a random variable X. Then this function is also a random variable, $g(X)$. We can find the expected value of $g(X)$ as follows.
For a discrete random variable,

$$E[g(X)] = \sum_i g(x_i)P\{X = x_i\}.$$ (2.31)

For a continuous random variable,

$$E[g(X)] = \int_{-\infty}^{\infty} g(x)f(x)dx.$$ (2.32)

The expected value only gives a single real value to describe a random variable, but the two random variables having the same expected value may exhibit different behavior of variability. *Variance*, denoted by $Var(X)$ or σ^2 of a random variable X, is measured as a squared distance of a real value from its expected value $E[X]$. A formula for $Var(X)$ is as follows.

$$Var(X) = E[(X - E[X])^2]$$ (2.33)

For a discrete random variable,

$$Var[X] = \sum_i (x_i - E[X])^2 P\{X = x_i\}.$$ (2.34)

For a continuous random variable,

$$Var[X] = \int_{-\infty}^{\infty} (x - E[X])^2 f(x)dx.$$ (2.35)

From (2.33), we can also find the variance from the following.

$$Var(X) = E[X^2] - 2E[XE[X]] + E[(E[X])^2]$$ (2.36)
$$= E[X^2] - 2E[X]E[X] + (E[X])^2$$
$$= E[X^2] - (E[X])^2$$

Variance is a measure of weighted deviations from the average value of a random variable. If the random variable assumes real values that move away from μ with high probability, the variance will be large. On the other hand, the variance will be small if the random variable assumes real values that lie closer to the average value μ.

The square root of a variance is called standard deviation, denoted by σ. We can also find the expected value of the jointly distributed random variables. Consider two random variables X and Y with a joint probability density function $f(x, y)$. We can calculate the expected value of a function $g(X, Y)$ of the two random variables as follows.

For discrete X and Y,

$$E[g(X, Y)] = \sum_i \sum_j g(x_i, y_j) P\{X = x_i, Y = y_j\}. \tag{2.37}$$

For continuous X and Y,

$$E[g(X, Y)] = \int_{-\infty}^{\infty} \int_{-\infty}^{\infty} g(x, y) f(x, y) dx dy. \tag{2.38}$$

If $g(X, Y) = aX + bY$, we can find the expectation from (2.40):

$$E[g(X, Y)] = \int_{-\infty}^{\infty} \int_{-\infty}^{\infty} (ax + by) f(x, y) dx dy \tag{2.39}$$

$$= \int_{-\infty}^{\infty} \int_{-\infty}^{\infty} axf(x, y) dx dy + \int_{-\infty}^{\infty} \int_{-\infty}^{\infty} byf(x, y) dx dy$$

$$= a \int_{-\infty}^{\infty} xf(x) dx + b \int_{-\infty}^{\infty} yf(y) dy$$

$$= aE[X] + bE[Y].$$

In general, for n random variables we can write (2.40):

$$E[a_1 X_1 + a_2 X_2 + \dots + a_n X_n] = a_1 E[X_1] + a_2 E[X_2] + \dots + a_n E[X_n]. \tag{2.40}$$

If $g(X, Y) = XY$, we have (2.41):

$$E[XY] = \int_{-\infty}^{\infty} \int_{-\infty}^{\infty} xyf(x, y) dx dy. \tag{2.41}$$

If X and Y are independent, then $f(x, y) = f(x)f(y)$, and the expectation will be as follows:

$$E[XY] = \int_{-\infty}^{\infty} \int_{-\infty}^{\infty} xyf(x)f(y) dx dy \tag{2.42}$$

$$= \left(\int_{-\infty}^{\infty} xf(x) dx \right) \left(\int_{-\infty}^{\infty} yf(y) dy \right)$$

$$= E[X]E[Y].$$

In the case of two random variables X and Y with a joint probability density function $f(x, y)$, we are also interested in seeing the deviations of the two random variables from their respective expected value. This measure is called *covariance,* denoted by $Cov(X, Y)$. We can find the covariance from the following expression:

$$Cov(X, Y) = E[(X - E[X])(Y - E[Y])]. \tag{2.43}$$

Equivalently,

$$\begin{aligned}
Cov(X, Y) &= E[(X - E[X])(Y - E[Y])] \\
&= E[XY - YE[X] - XE[Y] + E[X]E[Y]] \\
&= E[XY] - E[X]E[Y] - E[Y]E[X] + E[X]E[Y] \\
&= E[XY] - E[X]E[Y].
\end{aligned} \tag{2.44}$$

Note that if X and Y are independent, $E[XY] = E[X]E[Y]$, and $Cov(X, Y) = 0$. For jointly discrete random variables,

$$Cov(X, Y) = \sum_i \sum_j (x_i - E[X])(y_j - E[Y])P\{X = x_i, Y = y_j\}. \tag{2.45}$$

For jointly continuous random variables,

$$Cov(X, Y) = \int_{-\infty}^{\infty} \int_{-\infty}^{\infty} (x - E[X])(y - E[Y])f(x, y)dxdy. \tag{2.46}$$

The covariance gives the tendency that the two variables will vary together. This means that if X and Y move in the same direction, both deviations will be in the same sign, either positive or negative, and the resulting covariance will be positive. If X and Y vary in the opposite direction, then the deviation of each variable will be in different sign, and the covariance will be negative. However, when X and Y are independent, the covariance will be zero. It should also be noted from (2.33) that $Var(X) = E[(X - E[X])^2]$; this means that $Var(X) = Cov(X, X)$.

For any random variables X, Y and Z we can also write

$$\begin{aligned}
Cov(X, Y + Z) &= E[X(Y + Z)] - E[X]E[Y + Z] \\
&= E[XY] + E[XZ] - E[X]E[Y] - E[X]E[Z] \\
&= Cov(X, Y) + Cov(X, Z).
\end{aligned} \tag{2.47}$$

We can use this property to calculate variance of sum of random variables X_i as follows:

$$\begin{aligned}
Var\left(\sum_i X_i\right) &= Cov\left(\sum_i X_i, \sum_j X_j\right) \\
&= \sum_i \sum_j Cov(X_i, X_j) \\
&= \sum_i Cov(X_i, X_i) + 2\sum_i \sum_{j<i} Cov(X_i, X_j) \\
&= \sum_i Var(X_i) + 2\sum_i \sum_{j<i} Cov(X_i, X_j).
\end{aligned} \tag{2.48}$$

Note that when all X_i are independent, then $Cov(X_i, X_j) = 0$, and $Var(\sum_i X_i) = \sum_i Var(X_i)$.

We can use another dimensionless quantity called *correlation coefficient* to measure the tendency of two random variables. The correlation coefficient, denoted $Corr(X, Y)$ or $\rho_{X,Y}$, is defined below:

$$Corr(X, Y) = \frac{Cov(X, Y)}{\sqrt{(Var(X)Var(Y))}} = \frac{Cov(X, Y)}{\sigma_X \sigma_Y}. \tag{2.49}$$

It can be shown from Cauchy-Schwarz inequality that the correlation coefficient lies in the range $[-1, 1]$. The correlation coefficient can be thought of as the covariance of two random variables being normalized by the product of standard deviation of the two random variables. The correlation coefficient can only indicate linear dependence of two random variables. If the two random variables are independent, the covariance will be zero and the correlation coefficient will also be zero. However, the reverse is not true i.e. if the correlation coefficient is zero, it does not imply that the two random variables are independent.

2.7 Moment Generating Function

It can be seen that the expectation $E[X]$ and variance, $Var(X) = E[X^2] - (E[X])^2$, of a random variable can be computed from the expected value of a simple function $g(X) = X^k$ of a random variable when $k = 1$ and 2. The expectation of this function $g(X)$ is called k^{th} *initial* or *raw moment* of X, denoted by μ_k and is given by

$$\mu_k = E[X^k]. \tag{2.50}$$

For a discrete random variable X, we have

$$\mu_k = \sum_i x_i^k P\{X = x_i\}. \tag{2.51}$$

For a continuous random variable X having density function $f(x)$, we have

$$\mu_k = \int_{-\infty}^{\infty} x^k f(x) dx. \tag{2.52}$$

The first initial moment, $\mu_1 = E[X]$, is the expected value of a random variable. Similarly, we define the kth *central moment* of X, denoted μ'_k, as (2.53):

$$\mu'_k = E[(X - E[X])^k]. \tag{2.53}$$

For a discrete random variable X, we have (2.54):

$$\mu'_k = \sum_i (x_i - E[X])^k P\{X = x_i\}. \tag{2.54}$$

For a continuous random variable X having a density function $f(x)$, we have (2.55):

$$\mu'_k = \int_{-\infty}^{\infty} (x - E[X])^k f(x) dx. \tag{2.55}$$

The second central moment, $\mu'_2 = E[(X - E[X])^2]$, is the variance of a random variable.

If a random variable is symmetrical around its expected value, the odd central moments will be zero. The effect of asymmetry of the distribution can be detected from the odd central moments and is assessed by the following expression, called *skewness*:

$$Skew = \frac{\mu'_3}{\sqrt{\mu'^3_2}}. \tag{2.56}$$

We can convert the raw moment to central moment and vice versa. We can also find a moment of two or more random variables $X_i, i = 1, 2, \ldots, n$ in a similar manner:

$$\mu(k_1, k_2, \ldots, k_n) = E\left[X_1^{k_1} X_2^{k_2} \ldots X_n^{k_n}\right] \tag{2.57}$$

$$\mu'(k_1, k_2, \ldots, k_n) = E[(X_1 - E[X_1])^{k_1} (X_2 - E[X_2])^{k_2} \ldots (X_n - E[X_n])^{k_n}].$$

The moment is a powerful tool used to match two distributions. It can also be used for fitting a distribution to the raw data or when approximating a discrete distribution with a continuous distribution. We can generate moments of a random variable X using a *moment generating function*, denoted by $\phi(t)$.

The moment generating function is defined as:

$$\phi(t) = E[e^{tX}]. \tag{2.58}$$

Let us differentiate this function one time with respect to t,

$$\phi'(t) = \frac{d}{dt}\phi(t) = \frac{d}{dt}E[e^{tX}] = E\left[\frac{d}{dt}e^{tX}\right] = E[Xe^{tX}], \tag{2.59}$$

and then differentiate this function one more time with respect to t,

$$\phi''(t) = \frac{d^2}{dt^2}\phi(t) = \frac{d}{dt}E[Xe^{tX}] = E\left[\frac{d}{dt}Xe^{tX}\right] = E[X^2 e^{tX}]. \tag{2.60}$$

If we let $t = 0$, we have $\phi'(0) = E[X]$, and $\phi''(0) = E[X^2]$. This means that the moment generating function allows us to simply calculate the successive moments by differentiating the moment generating function and substitute $t = 0$.

Generally,

$$E[X^k] = \phi^{(k)}(0). \tag{2.61}$$

We can use the moment generating function to calculate the moments of summation of two independent random variables, $X + Y$. The moment generating function of this summation is given by (2.62):

$$\phi_{X+Y}(t) = E[e^{t(X+Y)}] = E[e^{tX}e^{tY}] = E[e^{tX}]E[e^{tY}] = \phi_X(t)\phi_Y(t). \tag{2.62}$$

In general, the moment generating function of the summation of independent random variables, $X_i, i = 1, 2, \ldots, n \ldots, n$, can be found as follows. Let $Y = \sum_i X_i$; then

$$\phi_Y(t) = E[e^{t\sum_i X_i}] = E\left[\prod_i e^{tX_i}\right] = \prod_i E[e^{tX_i}] = \prod_i \phi_{X_i}(t). \tag{2.63}$$

It is important to note that the moment generating function uniquely determines the distribution function.

In reliability analysis, we often deal with a random variable X representing time. This means that the random variable will only assume value from zero to infinity. In this case, we can calculate its moments from its density function, $f(x)$, using Laplace transformation. The Laplace transformation of the density function is denoted $\bar{f}(s)$, where s is a complex variable:

$$L[f(x)] = \bar{f}(s) = \int_0^\infty f(x)e^{-sx}dx = E[e^{-sX}]. \tag{2.64}$$

This expression is similar to the definition of the moment generating function shown in (2.58). The only difference is the negative sign of the variable s.

We can use Laplace transformation of the density function to calculate for the kth initial moments from

$$E[X^k] = (-1)^k \bar{f}^{(k)}(0). \tag{2.65}$$

We can also use the moments to help us construct a distribution function. Using Taylor's expansion to (2.64) and $e = \sum_{k=0}^\infty \frac{a^k}{k!}$, we have:

$$\bar{f}(s) = E\left[\sum_{k=0}^\infty \frac{(-sX)^k}{k!}\right] \tag{2.66}$$

$$= E\left[\sum_{k=0}^\infty \frac{(-1)^k s^k X^k}{k!}\right]$$

$$= \sum_{k=0}^\infty (-1)^k \frac{s^k}{k!} E[X^k]$$

$$= \sum_{k=0}^\infty (-1)^k \frac{s^k}{k!} \mu_k.$$

If we have information about the moments of a random variable, we can use (2.66) to construct a distribution function.

2.8 Functions of Random Variables

We examine in this section some random variables, both discrete and continuous, that are commonly used in power system reliability applications.

2.8.1 Bernoulli Random Variable

A Bernoulli random variable, X, is a discrete random variable whose outcome can only be either success or failure. This is also called a Bernoulli trial. In power system reliability analysis we usually use this distribution to represent a status of a transmission line, which can be either up or down. This can be denoted by a discrete random variable assuming 0 if it is down and 1 if it is up. The following probability mass function is used to characterize a Bernoulli random variable:

$$P\{X = 0\} = 1 - p \tag{2.67}$$

$$P\{X = 1\} = p, \tag{2.68}$$

where probability p is between 0 and 1, which denotes probability of success.

It can be seen that this distribution is only concerned with two possible outcomes of a component. The next distribution considers outcomes of multiple components.

2.8.2 Binomial Random Variable

Consider a generation system with n identical generators, and each generator is working independently and has probability of working (success) of p, thereby having a failure probability of $1 - p$. We are interested in knowing the number of unit(s) that is (are) working. Let X be number of working (success) unit(s) among n generators, taking value of $0, 1, 2, \dots, n$.

A random variable X is said to have binomial distribution with parameter (n, p). The probability mass function of this discrete random variable is given by (2.69):

$$P\{X = a\} = \binom{n}{a} p^a (1 - p)^{n-a}, \tag{2.69}$$

where $\binom{n}{a} = \frac{n!}{a!(n-a)!}$.

The expected value of X, $E[X] = np$, and variance is $Var[X] = np(1 - p)$.

Example 2.8 For a system of three identical and independent generators, each having probability of success of 0.9, let us find the probability that two generators are working.

In this example, X is a binomial random variable with parameter $(3, 0.9)$. The probability that two generators are working can be found as follows:

$$P\{X = 2\} = \binom{3}{2} 0.9^2(1 - 0.9) = 0.243.$$

In general, binomial distribution is used to describe n independent trials, each trail resulting in either success with probability p or failure with probability $1 - p$. Then, a binomial random variable, X, denotes the number of success in n independent trials.

Consider a case when number of trials reaches a very large quantity, and the probability of success is small. Let Λ be number of successes in n independent trials. We can approximate the success probability by $p = \frac{\Lambda}{n}$. Then,

$$P\{X = a\} = \binom{n}{a} p^a(1 - p)^{n-a} \tag{2.70}$$

$$= \frac{n!}{a!(n - a)!} \left(\frac{\Lambda}{n}\right)^a \left(1 - \frac{\Lambda}{n}\right)^{n-a}.$$

As $n \to \infty$, using binomial series, $(1 + \gamma)^\alpha = \sum_{k=0}^{\infty} \binom{\alpha}{k}\gamma^k$, and Taylor's expansion, $e^\alpha = \sum_{k=0}^{\infty} \frac{\alpha^k}{k!}$, we have the following approximation:

$$\left(1 - \frac{\Lambda}{n}\right)^{n-a} \approx \left(1 - \frac{\Lambda}{n}\right)^n \tag{2.71}$$

$$= 1 + \binom{n}{1}\frac{-\Lambda}{n} + \binom{n}{2}\left(\frac{-\Lambda}{n}\right)^2 + \dots$$

$$\approx 1 - \Lambda + \frac{\Lambda^2}{2!} - \frac{\Lambda^3}{3!} + \dots$$

$$= e^{-\Lambda}.$$

Then, the probability is written as (2.72):

$$P\{X = a\} = \frac{(n(n - 1) \dots (n - a + 1)(n - a)!)}{a!(n - a)!} \frac{\Lambda^a}{n^a} e^{-\Lambda} \tag{2.72}$$

$$= \frac{(n(n - 1) \dots (n - a + 1))}{a!} \frac{\Lambda^a}{n^a} e^{-\Lambda}.$$

As $n \to \infty$, $n(n - 1) \dots (n - a + 1) \approx n^a$, we have (2.73):

$$P\{X = a\} = \frac{\Lambda^a}{a!} e^{-\Lambda}. \tag{2.73}$$

Expression (2.73) shows that when a number of trials is very large, we can calculate the probability that the trial will be successful a times by using the average number of successes, Λ, over a long-run trial. This leads us to the next distribution, called Poisson distribution.

2.8.3 Poisson Random Variable

A Poisson random variable X is a discrete random variable taking value $0, 1, 2, \ldots$ with a parameter Λ, for some $\Lambda > 0$. The probability mass function of this random variable is given in (2.74):

$$P\{X = a\} = \frac{\Lambda^a}{a!} e^{-\Lambda}. \tag{2.74}$$

The expected value of X, $E[X] = \Lambda$ and variance is $Var[X] = \Lambda$.

A Poisson random variable is widely used to describe the number of occurrences (either failures or successes) in a fixed time given that an average or expected number of occurrences is Λ. For example, in reliability analysis it is commonly used to describe number of failures of a component within a certain time period given that the number of failures on average, Λ, is known.

Example 2.9 A transmission line fails on an average two times per year. If the number of failures can be described by Poisson distribution, what is the probability of having two failures in 5 years? What is the probability of having three failures in 10 years?

We first let X_5 be a Poisson random variable representing the number of failures in 5 years; its expected or average number of occurrences is $\Lambda_5 = 2 \times 5 = 10$ failures in 5 years. Then, the probability mass function of this random variable is

$$P\{X_5 = a\} = \frac{10^a}{a!} e^{-10}.$$

The probability of having two failures in 5 years is $P\{X_5 = 2\} = \frac{10^2}{2!} e^{-10} = 0.00227$.

Similarly, let X_{10} be a Poisson random variable representing number of failures in 10 years; its average number of occurrences is $\Lambda_{10} = 2 \times 10 = 20$ failures in 10 years. Then, the probability mass function of this random variable is

$$P\{X_{10} = a\} = \frac{20^a}{a!} e^{-20}.$$

The probability of having three failures in 5 years is $P\{X_{10} = 3\} = \frac{20^3}{3!} e^{-20} = 4.12 \times 10^{-7}$.

If, after a long experiment, we found out that within a time period, t, we have a number of failures equal to Λ; the average number of failures is $\lambda = \frac{\Lambda}{t}$; this number is the average number of occurrences per unit time interval. The probability mass function can be rewritten as (2.75):

$$P\{X = a\} = \frac{(\lambda t)^a}{a!} e^{-\lambda t}.$$

(2.75)

If number of failure in this time interval is zero, then

$$P\{X = 0\} = e^{-\lambda}.$$

(2.76)

We will see later in this chapter that (2.76) gives a widely known distribution function of a continuous random variable called exponential.

2.8.4 Uniform Random Variable

A continuous random variable is uniformly distributed on interval (α, β) if the probability density function is given as (2.77):

$$f(x) = \begin{cases} \frac{1}{\beta - \alpha} & \text{if } \beta < x < \alpha \\ 0 & \text{otherwise.} \end{cases}$$

(2.77)

The probability distribution function is given as (2.78):

$$f(x) = \begin{cases} 0 & \text{if } a \leq \beta \\ \frac{a - \alpha}{\beta - \alpha} & \text{if } \beta < a < \alpha \\ 1 & \text{if } a \geq \beta. \end{cases}$$

(2.78)

This distribution is frequently used for generating random numbers for Monte Carlo simulation.

2.8.5 Exponential Random Variable

An exponential random variable, X, is a non negative random variable on $[0, \infty)$ whose probability density function is given by (2.79) for some positive constant, $\lambda > 0$:

$$f(x) = \lambda e^{-\lambda x}$$

(2.79)

The probability distribution function is given in (2.80):

$$F(a) = 1 - e^{-\lambda a}.$$

(2.80)

If we take Laplace transformation as follows, we can find the moments s_j from the following moment generating function (2.81):

$$E[X^k] = (-1)^k \bar{f}^{(k)}(0),$$

(2.81)

where

$$L[f(x)] = \bar{f}(s) = \int_0^\infty f(x)e^{-sx}dx = \frac{\lambda}{\lambda + s}.$$ (2.82)

The expected value is the first moment given by (2.83):

$$E[X] = -\frac{d\bar{f}(s)}{ds}\Big|_{s=0} = \frac{\lambda}{(\lambda + s)^2}\Big|_{s=0} = \frac{1}{\lambda}.$$ (2.83)

Similarly, the second moment is given by (2.84):

$$E[X^2] = -\frac{d^2\bar{f}(s)}{ds}\Big|_{s=0} = \frac{2\lambda}{(\lambda + s)^3}\Big|_{s=0} = \frac{2}{\lambda^2}.$$ (2.84)

Then, the variance is found from (2.85):

$$Var(X) = E[X^2] - (E[X])^2 = \frac{1}{\lambda^2}.$$ (2.85)

If X denotes time to failure of a component, the expected value of X gives the expected value of time to failure of the component, and λ is called failure rate of the component. The survival function is shown in (2.86):

$$R(a) = \int_a^\infty f(x)dx = e^{-\lambda a}.$$ (2.86)

The hazard rate function can be found using (2.20),

$$h(a) = \frac{f(a)}{R(a)} = \lambda.$$ (2.87)

Recall that the hazard rate function gives the rate of a conditional probability of failure at time a given that the component has survived up to time a. As $\Delta a \to 0$, the hazard rate function can be written as $h(a)\Delta a = P\{a \leq X \leq a + \Delta a | X > a\}$. When the hazard rate function is constant, it implies that a component will fail at a constant rate irrespective of how long it has been in operation. This property is known as *memoryless property* and can be proven by finding the residual lifetime of a component.

When X is a random variable denoting time to failure of a component, assume that a component has operated up to time t without failure. Let $Y = X - t$ denote residual lifetime of this component. Then,

$$F_Y(a) = Pr\{Y \leq a | X > t\}.$$ (2.88)

The equation (2.88) gives the distribution function $F_Y(a)$ of a random variable Y, representing a residual lifetime of this component. Using conditional probability rule, we can find

$$F_Y(a) = Pr\{Y \leq a | X > t\} \tag{2.89}$$
$$= Pr\{X - t \leq a | X > t\}$$
$$= \frac{Pr\{t < X \leq a + t\}}{Pr\{X > t\}}$$
$$= \frac{\int_t^{a+t} \lambda e^{-\lambda x} dx}{e^{-\lambda t}}$$
$$= 1 - e^{-\lambda a}.$$

Since X is exponentially distributed, $F_X(a)$ is given by (2.80), which is exactly the same as above. This means that Y, the residual lifetime, is also exponentially distributed. The distribution of residual lifetime is independent of the time that a component has been in operation. It will not fail because of the gradual degradation of the component itself but from the random failure; in other words, this component does not age. Since the hazard rate function uniquely determines the distribution function, it follows from (2.87) that the exponential distribution is the only distribution that has memoryless property.

There is a connection between Poisson distribution and exponential distribution. Recall that Poisson distribution is used to describe the number of occurrences (either failures or successes) in a fixed time t given that an average or expected number of occurrences is Λ. Let X denotes = number of failures in this time interval $(0, t)$, and X is a Poisson random variable. The average number of failures within this time interval is $\lambda = \frac{\Lambda}{t}$. We found out from (2.76) that when number of failures is zero during this time interval, the probability is $P\{X = 0\} = e^{-\lambda t}$. In other words, a system has been operated in the time $(0, t)$ and no failure happens, which implies that a system will fail at a time greater than t.

Let Y denote time to failure of this system; then $Pr\{Y > t\} = e^{-\lambda t}$, and we can write

$$F_Y(t) = Pr\{Y \leq t\} = 1 - Pr\{Y > t\} = 1 - e^{-\lambda t}.$$

This shows that the random variable Y is exponentially distributed. Thus, we can conclude that for a Poisson random variable X, if we focus on time between failures and denote this time by a continuous random variable Y, this time is exponentially distributed.

2.8.6 Normal Random Variable

A normal random variable, X, is a continuous random variable on $(-\infty, \infty)$ with parameter μ and σ^2, whose probability density function is given by (2.90). We can write $X \sim N(\mu, \sigma^2)$.

$$f(x) = \frac{1}{\sigma\sqrt{2\pi}} e^{\frac{-(x-\mu)^2}{2\sigma^2}}, x \in (-\infty, \infty) \tag{2.90}$$

The normal distribution is bell-shaped and has symmetry around μ, which is its mean value. The variance of this random variable is $Var[X] = \sigma^2$. The probability distribution function is given by (2.91):

$$F(x) = \frac{1}{\sigma\sqrt{2\pi}} \int_{-\infty}^{x} e^{\frac{-(x-\mu)^2}{2\sigma^2}} dx. \tag{2.91}$$

Let $z = \frac{(x-\mu)}{\sigma}$, $dx = \sigma dz$. We can write (2.92):

$$F(x) = \frac{1}{2\pi} \int_{-\infty}^{\frac{x-\mu}{\sigma}} e^{-\frac{z^2}{2}} dz. \tag{2.92}$$

The integral in (2.92) cannot be expressed explicitly. Numerical integration is used to obtain the value of this integral in a tabular form.

We can also use the moment generating function to calculate for mean and variance as follows:

$$\phi(t) = E[e^{tX}] = \frac{1}{\sigma\sqrt{2\pi}} \int_{-\infty}^{\infty} e^{tx - \frac{(x-\mu)^2}{2\sigma^2}} dx. \tag{2.93}$$

Let $z = \frac{x-\mu}{\sigma}$ and $dx = \sigma dz$. We can write

$$\phi(t) = \frac{1}{\sigma\sqrt{2\pi}} \int_{-\infty}^{\infty} e^{t(\sigma z + \mu) - \frac{z^2}{2}} dz$$

$$= \frac{e^{t\mu}}{\sqrt{2\pi}} \int_{-\infty}^{\infty} e^{-\frac{(z-\sigma t)^2}{2} + \frac{\sigma^2 t^2}{2}} dz$$

$$= e^{t\mu + \frac{\sigma^2 t^2}{2}} \frac{1}{\sqrt{2\pi}} \int_{-\infty}^{\infty} e^{-\frac{(z-\sigma t)^2}{2}} dz.$$

Let $w = z - \sigma t$ and $dw = dz$. We have

$$\phi(t) = e^{t\mu + \frac{\sigma^2 t^2}{2}} \frac{1}{\sqrt{2\pi}} \int_{-\infty}^{\infty} e^{-\frac{w^2}{2}} dw = e^{t\mu + \frac{\sigma^2 t^2}{2}}. \tag{2.94}$$

The first and second moments are found from $\phi'(0) = E[X] = \mu$ and $\phi''(0) = E[X^2] = \mu^2 + \sigma^2$. The mean and variance are the same as previously stated.

This function leads to the special case of normal random variable when the mean value is zero and variance is one, $Z \sim N(0, \sigma^2)$. This random variable Z is called standard normal random variable, which has the following density function:

$$f(z) = \frac{1}{\sqrt{2\pi}} e^{-\frac{z^2}{2}}, z \in (-\infty, \infty). \tag{2.95}$$

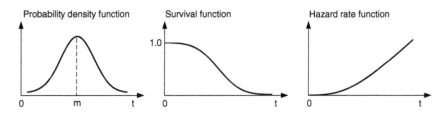

Figure 2.16 Characteristics of a normal random variable.

It should be noted that normal random variable can take any value in the real axis. When we use this random variable, $X \sim N(\mu, \sigma^2)$, to represent operating time of a component, the function needs to be modified to reflect the fact that the operation time can only be positive. This is done by truncating the normal distribution by (2.96):

$$f(x) = \frac{1}{\alpha \sigma \sqrt{2\pi}} e^{-\frac{(x-\mu)^2}{2\sigma^2}}, x \in [0, \infty), \tag{2.96}$$

where α is a normalizing parameter and $\alpha = \frac{1}{\sigma\sqrt{2\pi}} \int_0^\infty e^{-\frac{(x-\mu)^2}{2\sigma^2}} dx$.

The survival function and hazard rate function can be found from (2.16) and (2.20). It should be noted that the hazard rate function of the normal distribution is monotonically increasing, as shown in Figure 2.16.

2.8.7 Log-Normal Random Variable

A log-normal random variable, X, is a continuous random variable on $(0, \infty)$ with parameter μ and σ^2, and if $Y = \log X$ is normally distributed with parameter μ and σ^2. Since a logarithm is a non decreasing function, we can write

$$F_X(x) = Pr\{X \le x\} = Pr\{\log X \le \log x\} = Pr\{Y \le y\} = F_Y(y).$$

Using chain rule, we have

$$f_X(x) = \frac{d}{dx} F_X(x)$$

$$= \frac{d}{dx} F_Y(y)$$

$$= f_Y(y) \frac{dy}{dx}$$

$$= f_Y(\log x) \frac{dy}{dx}$$

$$= \frac{1}{x} f_Y(\log x).$$

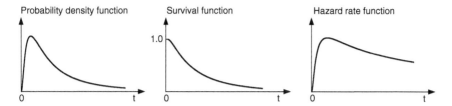

Figure 2.17 Characteristics of a log-normal random variable.

The probability density function of X is (2.97):

$$f(x) = \frac{1}{x\sigma\sqrt{2\pi}} e^{-\frac{(\log x - \mu)^2}{2\sigma^2}}, x \in (0, \infty). \tag{2.97}$$

The mean and variance of this random variable can be found from moment generating function. Since $Y = \log X$ is normally distributed with the moment generating function shown in (2.94), and $X = e^Y$, we can write (2.98):

$$E[X] = E[e^Y]. \tag{2.98}$$

From $\phi_Y(t) = E[e^{tY}] = e^{t\mu + \frac{\sigma^2 t^2}{2}}$, let $t = 1$, and we have $E[X] = e^{\mu + \frac{\sigma^2}{2}}$, the mean value of X. Similarly, $E[X^2] = E[e^{2Y}]$, let $t = 2$, and we have $E[X^2] = e^{2\mu + 2\sigma^2}$. The variance can be found from (2.99):

$$Var(X) = E[X^2] - (E[X])^2 = e^{2(\mu + \sigma^2)} - e^{\mu + \sigma^2}. \tag{2.99}$$

The hazard rate function of log-normal is shown in Figure 2.17, which demonstrates that the function is not monotonically increasing and does not seem to model a lifetime of a physical component. However, in several cases, it shows reasonable fit for repair times.

2.8.8 Gamma Random Variable

A gamma random variable, X, is a continuous random variable on $[0, \infty)$ with parameter $\lambda > 0$ and $\alpha > 0$, whose probability density function is given by (2.100):

$$f(x) = \frac{\lambda e^{-\lambda x}(\lambda x)^{\alpha-1}}{\Gamma(\alpha)}, x \in [0, \infty). \tag{2.100}$$

where $\Gamma(\alpha)$ is a gamma function and is defined as (2.101):

$$\Gamma(\alpha) \equiv \int_0^\infty z^{\alpha-1} e^{-z} dz. \tag{2.101}$$

Note that when α is a positive integer, we can write $\Gamma(\alpha) = (\alpha - 1)!$, thus $\Gamma(1) = \Gamma(2) = 1$; using integration by parts, we can write $\Gamma(\alpha + 1) = \alpha\Gamma(\alpha)$.

The probability distribution function is found from

$$F(a) = \frac{1}{\Gamma(\alpha)} \int_0^a \lambda e^{-\lambda x} (\lambda x)^{\alpha-1} dx.$$

Substitute $z = \lambda x$ and $dz = \lambda dx$ and we have (2.102):

$$F(a) = \frac{1}{\Gamma(\alpha)} \int_0^{\lambda a} e^{-z} z^{\alpha-1} dz. \tag{2.102}$$

When α is integer, we can use integration by parts to find the distribution function.

$$F(a) = 1 - \sum_{k=0}^{\alpha-1} \frac{\lambda e^{-\lambda a} (\lambda a)^k}{k!}, a \text{ is integer.} \tag{2.103}$$

We can compute the kth moment of X directly as follows:

$$E[X^k] = \int_0^{\infty} \frac{\lambda e^{-\lambda x} x^k (\lambda x)^{\alpha-1}}{\Gamma(\alpha)} dx.$$

Substitute $z = \lambda x$ and $dz = \lambda dx$ and we have

$$E[X^k] = \frac{1}{\lambda^k \Gamma(\alpha)} \int_0^{\infty} e^{-z} z^{k+\alpha-1} dz = \frac{\Gamma(k + \alpha)}{\lambda^k \Gamma(\alpha)}. \tag{2.104}$$

The first and second moment is found from the following:

$$E[X] = \frac{\Gamma(\alpha + 1)}{\lambda \Gamma(\alpha)} = \frac{\alpha \Gamma(\alpha)}{\lambda \Gamma(\alpha)} = \frac{\alpha}{\lambda} \tag{2.105}$$

$$E[X^2] = \frac{\Gamma(\alpha + 2)}{\lambda^2 \Gamma(\alpha)} = \frac{\alpha^2}{\lambda^2}. \tag{2.106}$$

The mean value is $\frac{\alpha}{\lambda}$, and the variance of this random variable is $Var[X] = \frac{\alpha}{\lambda^2}$.

The survival function is shown in (2.107):

$$R(a) = P\{X > a\} = \int_a^{\infty} \frac{\lambda e^{-\lambda x} (\lambda x)^{\alpha-1}}{\Gamma(\alpha)} dx. \tag{2.107}$$

The hazard rate function of a gamma random variable is

$$h(a) = \frac{f(a)}{R(a)} = \frac{1}{\int_a^{\infty} e^{-\lambda(x-a)} (\frac{x}{a})^{\alpha-1} dx}.$$

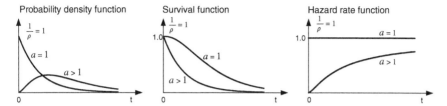

Probability density function   Survival function   Hazard rate function

Figure 2.18 Characteristics of a gamma random variable.

Let $z = x - a$ and $dz = dx$; then,

$$h(a) = \frac{1}{\int_0^\infty e^{-\lambda z}\left(1 + \frac{z}{a}\right)^{\alpha-1} dz}. \tag{2.108}$$

Note that if $\alpha = 1$, then $\int_0^\infty e^{-\lambda z} dz = \lambda$, which is a constant value. If $\alpha > 1$, the hazard rate function will be increasing; if $0 < \alpha < 1$, the hazard rate function will be decreasing, as shown in Figure 2.18.

2.8.9 Weibull Random Variable

A Weibull random variable, X, is a continuous random variable on $(0, \infty)$ with parameter $\lambda > 0$ and $\alpha > 0$, and if Y is exponentially distributed with parameter λ, then $X = Y^{\frac{1}{\alpha}}$ or $Y = X^\alpha$. This means that an exponential random variable is a special case of a Weibull random variable with $\alpha = 1$. Since $\alpha > 0$, we can write

$$F_X(x) = Pr\{X \leq x\} = Pr\{X^\alpha \leq x^\alpha\} = Pr\{Y \leq y\} = F_Y(y).$$

Thus, the probability distribution function is (2.109):

$$F_X(x) = 1 - e^{-\lambda y} = 1 - e^{-\lambda x^\alpha}, x \in (0, \infty). \tag{2.109}$$

Using chain rule, we have

$$f_X(x) = \frac{d}{dx} F_X(x)$$

$$= \frac{d}{dx} F_Y(y)$$

$$= f_Y(y)\frac{dy}{dx}$$

$$= f_Y(x^\alpha)\frac{dy}{dx}$$

$$= \alpha x^{\alpha-1} f_Y(x^\alpha).$$

The probability density function of X is (2.110), which can also be found from differentiating (2.109):

$$f(x) = \alpha \lambda x^{\alpha-1} e^{-\lambda x^{\alpha}}, x \in (0, \infty). \tag{2.110}$$

For a Weibull random variable, it is simpler to compute the kth moment of X directly as follows:

$$E[X^k] = \int_0^{\infty} x^k \alpha \lambda x^{\alpha-1} e^{-\lambda x^{\alpha}} dx.$$

Substitute $z = \lambda x^{\alpha}$ and $dz = \alpha \lambda x^{\alpha-1} dx$, and we have

$$E[X^k] = \frac{1}{\lambda^{\frac{k}{\alpha}}} \int_0^{\infty} z^{\frac{k}{\alpha}} e^{-z} dz.$$

From the previous section, $\Gamma(a) = \int_0^{\infty} z^{a-1} e^{-z} dz$; we can use gamma function to calculate the moment as follows:

$$E[X^k] = \frac{1}{\lambda^{\frac{k}{\alpha}}} \Gamma\left(1 + \frac{k}{\alpha}\right). \tag{2.111}$$

The expected value is found when $k = 1$,

$$E[X] = \frac{1}{\lambda^{\frac{1}{\alpha}}} \Gamma\left(1 + \frac{1}{\alpha}\right). \tag{2.112}$$

The second moment is when $k = 2$, $E[X^2] = \frac{1}{\lambda^{\frac{2}{\alpha}}} \Gamma(1 + \frac{2}{\alpha})$. The variance is calculated directly from the first and second moment as follows:

$$Var[X] = \frac{1}{\lambda^{\frac{2}{\alpha}}} \Gamma(1 + 2/\alpha) - \left(\Gamma\left(1 + \frac{2}{\alpha}\right)\right)^2. \tag{2.113}$$

The survival function is shown in (2.114):

$$R(a) = P\{X > a\} = 1 - F(a) = e^{-\lambda a^{\alpha}}. \tag{2.114}$$

Thus, the hazard rate function is (2.115):

$$h(x) = \frac{f(x)}{R(x)} = \frac{\alpha \lambda x^{\alpha-1} e^{-\lambda x^{\alpha}}}{e^{-\lambda x^{\alpha}}} = \alpha \lambda x^{\alpha-1}. \tag{2.115}$$

Expression (2.115) shows that if $\alpha = 1$, the hazard rate function is constant. If $\alpha > 1$, the hazard rate function is increasing, and if $\alpha < 1$, the hazard rate function is decreasing. With this flexibility, a Weibull random variable is often used to model time to failure of a component. The characteristics of this distribution are given in Figure 2.19.

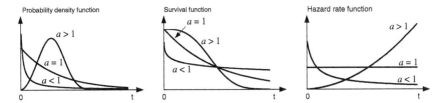

Probability density function   Survival function   Hazard rate function

Figure 2.19 Characteristics of a Weibull random variable.

Exercises

2.1 If the number of faults occurring on a transmission line is a Poisson random variable with an average of three faults per year, what is the probability that no fault occurs in two years? What is the probability that at least two faults occur in five years?

2.2 A transformer's lifetime, t (year), is assumed to have a uniform distribution, where $f(t) = a, 0 < t \leq 4$ and $f(t) = 0$ otherwise. What is the hazard rate function of this transformer? What is the probability that this transformer will fail in the period between zero and one year of operation?

3

Review of Stochastic Process

3.1 Introduction

We present the basic concepts of stochastic process in this chapter. The main focus is on discrete-time and continuous-time Markov processes with an application to power systems reliability.

Consider a system called system A, for example, of two transmission lines. In the previous chapter, we used discrete random variables to describe the status of the two transmission lines, as shown below in Figure 3.1.

We can observe the status of the two transmission lines over a period of time and plot the value of this discrete random variable as a function of time, as in Figure 3.2.

We can also observe another identical system, called system B, of two transmission lines operated independently from system A. Since both transmission lines in the two systems are identical, we may observe the status of the two transmission lines in system B as shown in Figure 3.3

At each hour, the status of the two transmission lines from both systems can be 0, 1 or 2 depending on their random behavior. Therefore, the status of the two transmission lines at a snapshot in time is a random variable. When the random variable in the system evolves with time, it is called a random process or stochastic process. An observation of the system is called a realization of the stochastic process. A realization can be different from others due to randomness in the process. For the previous two transmission line systems A and B, Figure 3.2 and 3.3 represent two independent realizations of the system. The realization of an identical and independent process provides an insight on how the system develops through time.

In general, a stochastic process is a collection of random variables $\{Z(t)\}, t \in T$, indexed by parameter t. The possible values that all random variables $Z(t)$, equivalently, $\{Z_t\}, t \in T$ can assume is called state of the system, and the set of all possible values of this random variable is called state space. The set T of all possible values of the indexing parameter is called the parameter space, it and usually represents time. When the index set T is discrete, we call it a

Electric Power Grid Reliability Evaluation: Models and Methods, First Edition. Chanan Singh, Panida Jirutitijaroen, and Joydeep Mitra.
© 2019 by The Institute of Electrical and Electronic Engineers, Inc. Published 2019 by John Wiley & Sons, Inc.

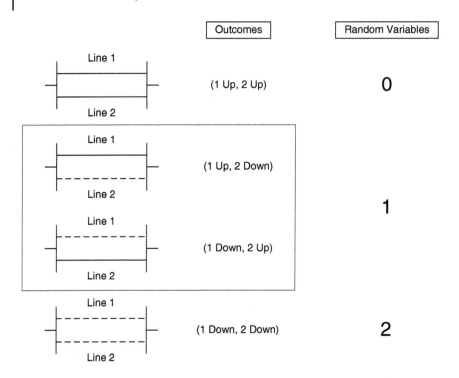

| Outcomes | Random Variables |

Figure 3.1 A status of two transmission lines.

discrete-time process, $\{Z_n\}$, $n = 0, 1, 2, \ldots$, and when the index set T is continuous, $\{Z_t\}$, $t \geq 0$, we call it a continuous-time process.

Stochastic processes can be classified by the nature of the random variable and parameter index into four categories as follows. In power system reliability

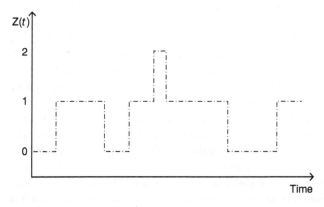

Figure 3.2 Observation of a status of two transmission lines in system A.

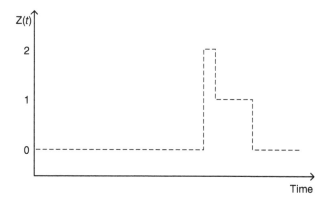

Figure 3.3 Observation of a status of two transmission lines in system B.

the main focus is on the discrete random variable in both continuous and discrete time.

1. Continuous random variable in a continuous-time process. For example, capacity of solar photovoltaic generated throughout a day, as the generating capacity can be any positive real value number, and the time is a continuous indexing parameter.
2. Continuous random variable in a discrete-time process. For example, an electric load observed at every hour.
3. Discrete random variable in a continuous-time process. For example, status of a device in the system throughout a year.
4. Discrete random variable in a discrete-time process. For example, status of a generator at the beginning of each hour.

In the previous chapter, we reviewed the concept of a random variable, which is a function that assigns numerical values to all outcomes in the state space. We also defined the probability density function that yields probabilities associated with all possible values of the random variable. In stochastic process, the random variable evolves with time, and the main goal is to be able to determine the *probability distribution* of $\{Z(t)\}, \forall t \in T$ at all times. For example, Figure 3.4 and 3.5 show the probability distribution of a discrete random variable in both discrete-time and continuous-time stochastic processes correspondingly. The primary interest is to find probability of Z_{n+1} in a discrete-time process or Z_t in a continuous-time process.

The straightforward calculation requires the knowledge of the joint probability distribution of $\{Z(t)\}, \forall t \in T$, which is rarely available in practice. Bayes' theorem can be used as an alternative to calculate probability distribution Z_t based on the concept of conditional probability. Consider an hourly measured electric load in power systems that changes over time. The current load value is likely to depend on the states where the process has been until now, which

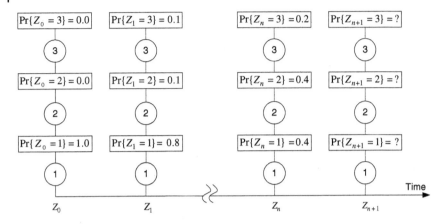

Figure 3.4 An example of a stochastic process with a discrete random variable and discrete time.

is given by an example of conditional probability of a discrete-time process as follows:

$$P(Z_{n+1} = j | Z_n = i, Z_{n-1} = k, \ldots, Z_0 = a_0). \tag{3.1}$$

We can simplify this expression with some assumptions on how the process develops over time. If the stochastic process is an *independent process*, then the conditional probability becomes

$$P(Z_{n+1} = j | Z_n = i, Z_{n-1} = k, \ldots, Z_0 = a_0) = P(Z_{n+1} = j). \tag{3.2}$$

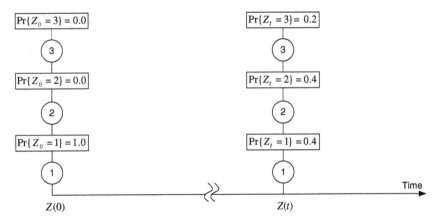

Figure 3.5 An example of a stochastic process with a discrete random variable and continuous time.

This simplification leads to the case where the current state of the process is independent of how the process has evolved. Even though it allows simple probability calculation, it is not realistic enough to represent physical systems due to the independence assumption. A balance between the dependency assumption to allow sufficient realism of the physical systems and the tractability of the probabilistic model should be met. A well-known class of the stochastic process called the *Markov process* offers an adequate balance by allowing a future state of Z_{n+1} of the stochastic process to depend only on the current state of Z_n, not the whole history. An example of the discrete-time Markov process is given below:

$$P(Z_{n+1} = j | Z_n = i, Z_{n-1} = k, \dots, Z_0 = a_0) = P(Z_{n+1} = j | Z_n = i). \quad (3.3)$$

The probability distribution of Z_{n+1} depends on the distribution of Z_n but not before. This property is called the *Markov property*. In case of a continuous-time Markov process, the Markov property is given as follows:

$$P(Z_{s+t} = j | Z_s = i, Z_u = a_u, \forall u \in [0, s)) = P(Z_{s+t} = j | Z_s = i). \quad (3.4)$$

The above expression reads: the probability of the random variable Z_{s+t} at time $s + t$ depends only on the present state of Z_s. Due to its dependency only on the present state, the Markov process is often referred to as a *memoryless* process.

If the conditional probability is constant over time, then the Markov process is called a *stationary process*. This property is called *time-homogeneous property*. For example, for a stationary discrete-time Markov process,

$$P(Z_{n+1} = j | Z_n = i) = P(Z_n = j | Z_{n-1} = i) == P(Z_1 = j | Z_0 = i). \quad (3.5)$$

For a stationary continuous-time Markov process, we have the following:

$$P(Z_{s+t} = j | Z_s = i) = P(Z_t = j | Z_0 = i). \quad (3.6)$$

The remainder of this chapter will focus on the discrete-time stationary Markov chain and continuous-time stationary Markov chain.

3.2 Discrete-Time Markov Process

The discrete-time Markov process is the stochastic process with discrete index set $\{Z_n, n = 0, 1, \dots, n\}$.

Recalling from the Markov property in (3.3), define P_{ij} as the transition probability from state i to state j,

$$P_{ij} = P(Z_{n+1} = j | Z_n = i). \quad (3.7)$$

For a stationary process, the transition probability is the same at any time step. For example,

$$P_{ij} = P(Z_{n+1} = j|Z_n = i) = P(Z_n = j|Z_{n-1} = i) = \ldots = P(Z_1 = j|Z_0 = i).$$

The transition probability P_{ij} gives the probability that the process will make a transition to state j, given that the process is at state i. Since this is a probability value, it follows that $0 \le P_{ij} \le 1$. The process must make a transition to some state in the state space, which implies that

$$\sum_{j=0}^{\infty} P_{ij} = 1, i = 0, 1, 2, \ldots \tag{3.8}$$

3.2.1 Transition Probability Matrix

We can write the transition probability between each state in the state space in a matrix form, called $\mathbf{P}$, or a transition probability matrix, as follows:

$$\mathbf{P} = [P_{ij}] = \begin{bmatrix} P_{00} & P_{01} & \cdots \\ P_{10} & P_{11} & \cdots \\ \vdots & \vdots & \vdots \end{bmatrix}. \tag{3.9}$$

This matrix gives the transition probability of a one-step transition from one state to others. Since the process is stationary, this transition probability matrix is constant. In engineering applications, a discrete-time homogeneous Markov process can be represented by a *state-transition diagram*, which shows all possible states as well as the transition probabilities. Figure 3.6 shows the state-transition diagram of a three-state system as an example.

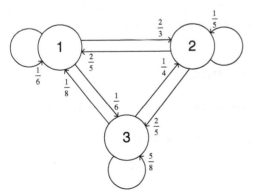

Figure 3.6 State-transition diagram of a three-state system.

Figure 3.7 State-transition diagram of Example 3.1.

Example 3.1   A person is doing target practice. If he misses the target, the probability of the next shot being a hit is $\frac{1}{2}$, but if he scores a hit the probability of the next shot being a hit is $\frac{3}{4}$. The process can be modeled as a two-state discrete-time Markov process. Draw a state-transition diagram and write a transition probability matrix of this process.

Denoting a hit and miss by states 0 and 1 respectively, the state-transition diagram is given in Figure 3.7.

The transition probability is given as $P_{00} = \frac{3}{4}$ and $P_{10} = \frac{1}{2}$. Then, the rest of transition probabilities can be found using (3.8), $P_{01} = 1 - P_{00} = \frac{1}{4}$, and $P_{11} = 1 - P_{10} = \frac{1}{2}$. The transition probability matrix is given as follows:

$$\mathbf{P} = \begin{bmatrix} \frac{3}{4} & \frac{1}{4} \\ \frac{1}{2} & \frac{1}{2} \end{bmatrix}.$$

We can also calculate transition probability after two time steps, denoted by $P_{ij}^{(2)}$, using one-step transition probability.

In the following, we illustrate the calculation of a three-state system. If the system has three states, 1, 2, and 3, the transition probability matrix of this system is given by

$$\mathbf{P} = \begin{bmatrix} P_{11} & P_{12} & P_{13} \\ P_{21} & P_{22} & P_{23} \\ P_{31} & P_{32} & P_{33} \end{bmatrix}.$$

We can compute the transition probability after two time steps from state 1 at time 0 to state 3 at time $n = 2$ as follows. When a process is in state 1, it can make a transition to state 3 in three ways through state 1 or state 2 or state 3 at time $n = 1$. This transition is depicted by Figure 3.8.

Using the conditional probability concept, the two-step transition probability from state 1 to state 3 can be written as follows:

$$P_{13}^{(2)} = P(Z_2 = 3, Z_1 = 1 | Z_0 = 1) + P(Z_2 = 3, Z_1 = 2 | Z_0 = 1)$$
$$+ P(Z_2 = 3, Z_1 = 3 | Z_0 = 1).$$

Since

$$P(Z_2 = 3, Z_1 = 1 | Z_0 = 1) = P(Z_2 = 3 | Z_1 = 1, Z_0 = 1) \times P(Z_1 = 1 | Z_0 = 1)$$
$$= P_{13} P_{11},$$

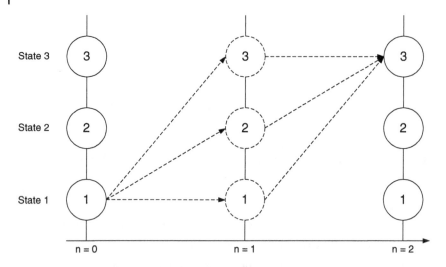

Figure 3.8 A transition from state 1 to state 3 of a three-state system.

by Markov and time-homogeneous properties, then

$$P_{13}^{(2)} = P_{11}P_{13} + P_{12}P_{23} + P_{13}P_{33}.$$

Similarly, all other transition probabilities can be found. Note that the above expression can be found by simply multiplying the first row of the transition probability matrix by the third column of the transition probability matrix. We can write two-step transition probabilities in the matrix form as follows:

$$\mathbf{P}^{(2)} = \begin{bmatrix} P_{11} & P_{12} & P_{13} \\ P_{21} & P_{22} & P_{23} \\ P_{31} & P_{32} & P_{33} \end{bmatrix} \times \begin{bmatrix} P_{11} & P_{12} & P_{13} \\ P_{21} & P_{22} & P_{23} \\ P_{31} & P_{32} & P_{33} \end{bmatrix} = \mathbf{P}^2.$$

In general, we define n-step transition probability matrix as $\mathbf{P}^{(n)}$, where each term in the matrix is defined as follows:

$$P_{ij}^{(n)} = P(Z_{n+m} = j | Z_m = i), n \geq 0, i,j \geq 0. \tag{3.10}$$

where $P_{ij}^{(1)} = P_{ij}$. Note that by definition, $\mathbf{P}^{(0)}$ is a transition probability matrix after 0 time step, i.e., the process does not make any transition, which means that the transition probability from state i to state j is

$$P_{ij}^{(0)} = P(Z_n = j | Z_n = i) = 0.$$

Since the process does not make any transition, the transition probability from state i to state i itself is

$$P_{ii}^{(0)} = P(Z_n = i | Z_n = i) = 1.$$

We can therefore conclude that $\mathbf{P}^{(0)} = \mathbf{I}$, which is an identity matrix.

Equation (3.10) expresses the transition probability of a process moving from state i to state j in n time steps. Similarly, we can write

$$P_{ij}^{(n+m)} = P(Z_{n+m} = j | Z_0 = i).$$

Assume that at time m, the process is at state k; then

$$P_{ij}^{(n+m)} = \sum_{k=0}^{\infty} P(Z_{n+m} = j, Z_m = k | Z_0 = i)$$

$$= \sum_{k=0}^{\infty} P(Z_{n+m} = j | Z_m = k, Z_0 = i) \times P(Z_m = k | Z_0 = i).$$

With Markov and time-homogeneous properties, we can write

$$P_{ij}^{(n+m)} = \sum_{k=0}^{\infty} P(Z_{n+m} = j | Z_m = k) P(Z_m = k | Z_0 = i).$$

This means that the transition probability of $n + m$ time steps is found by

$$P_{ij}^{(n+m)} = \sum_{k=0}^{\infty} P_{ik}^m P_{kj}^n. \tag{3.11}$$

Equation (3.11) is called the *Chapman-Kolmogorov equation*, which is used to calculate transition probability of n time-step transitions. Let $n = m = 1$. We have

$$P_{ij}^{(2)} = \sum_{k=0}^{\infty} P_{ik} P_{kj}.$$

Writing in a matrix form, $\mathbf{P}^{(2)} = \mathbf{P}^2$, in general, we can find a transition probability matrix after n time steps by (3.12):

$$\mathbf{P}^{(n)} = \mathbf{P}^n. \tag{3.12}$$

The transition probability only gives the probability that a process will make a transition from state i to state j at any time step.

3.2.2 Probability Distribution at Time Step n

The primary interest is also to find the probability distribution of the process, Z_n, at any time. Using transition probability concept, we can find the distribution given that the initial distribution at time 0, Z_0, is known. Define the probability distribution at initial time by (3.13):

$$\mathbf{p}^{(0)} = \left[p_0^{(0)}, p_1^{(0)}, \ldots, p_k^{(0)}, \ldots \right].$$

(3.13)

Note that $p_k^{(0)}$ is the probability that the process starts from state k at time 0, i.e., $p_k^{(0)} = P(Z_0 = k)$, and $\sum_{k=0}^{\infty} p_k^{(0)} = 1$.

Once the initial distribution and transition probabilities are known, the probability distribution of the process at any time step can be found by conditional probability (3.14):

$$P(Z_n = j) = \sum_{i=0}^{\infty} P(Z_n = j | Z_0 = i) \times P(Z_0 = i) = \sum_{i=0}^{\infty} P_{ij}^{(n)} p_i^{(0)}$$

(3.14)

Equation (3.14) gives the probability distribution at nth time step and can be written in a matrix form as follows:

$$\mathbf{p}^{(n)} = \mathbf{p}^{(0)} \mathbf{P}^{(n)}.$$

(3.15)

The probability distribution at any time step can now be calculated from (3.15). Since $\mathbf{P}^{(n)}$ is a square matrix, we can use matrix decomposition to write $\mathbf{P} = VDV^{-1}$ and $\mathbf{P}^n = VD^n V^{-1}$, where D is a diagonal matrix of eigen values and columns of V are corresponding eigen vectors. We can use this decomposition to find the time-specific distribution function of the process in a closed form.

Example 3.2 Continuing from Example 3.1, given that the initial shot is a hit, what is the probability of a hit on the 4th shot? Find probability distribution of this process at nth time step.

The process takes 3 steps to the 4th shot; the 3-step transition probability matrix is given by

$$\mathbf{P}^3 = \begin{bmatrix} \frac{3}{4} & \frac{1}{4} \\ \frac{1}{2} & \frac{1}{2} \end{bmatrix}^3 = \begin{bmatrix} \frac{43}{64} & \frac{21}{64} \\ \frac{21}{32} & \frac{11}{32} \end{bmatrix}.$$

Given that the initial shot is a hit, the initial probability distribution is $\mathbf{p}^{(0)} = \begin{bmatrix} 1 & 0 \end{bmatrix}$, and the probability of a hit at the 4th shot is, using (3.15),

$$\mathbf{p}^{(3)} = \mathbf{p}^{(0)} \mathbf{P}^3 = \begin{bmatrix} 1 & 0 \end{bmatrix} \times \begin{bmatrix} \frac{43}{64} & \frac{21}{64} \\ \frac{21}{32} & \frac{11}{32} \end{bmatrix}.$$

The probability of a hit on the 4th shot is $\frac{43}{64}$.

From $P = \begin{bmatrix} \frac{3}{4} & \frac{1}{4} \\ \frac{1}{2} & \frac{1}{2} \end{bmatrix}$, the eigen values are found from

$$det(\mathbf{P} - d\mathbf{I}) = \begin{vmatrix} \frac{3}{4} - d & \frac{1}{4} \\ \frac{1}{2} & \frac{1}{2} - d \end{vmatrix} = 0.$$

We have $d = 1, \frac{1}{4}$. We can then find the eigen vector for $d = 1$:

$$\begin{bmatrix} \frac{3}{4} - 1 & \frac{1}{4} \\ \frac{1}{2} & \frac{1}{2} - 1 \end{bmatrix} \begin{bmatrix} v_{11} \\ v_{21} \end{bmatrix} = \begin{bmatrix} 0 \\ 0 \end{bmatrix}.$$

Then $v_{11} = v_{21}$, let $v_{11} = 1$, $\begin{bmatrix} v_{11} \\ v_{21} \end{bmatrix} = \begin{bmatrix} 1 \\ 1 \end{bmatrix}$.
The eigen vector for $d = \frac{1}{4}$:

$$\begin{bmatrix} \frac{3}{4} - \frac{1}{4} & \frac{1}{4} \\ \frac{1}{2} & \frac{1}{2} - \frac{1}{4} \end{bmatrix} \begin{bmatrix} v_{12} \\ v_{22} \end{bmatrix} = \begin{bmatrix} 0 \\ 0 \end{bmatrix}.$$

Then $v_{12} = -\frac{1}{2}v_{22}$, let $v_{22} = 1$, $\begin{bmatrix} v_{12} \\ v_{22} \end{bmatrix} = \begin{bmatrix} -\frac{1}{2} \\ 1 \end{bmatrix}$.

Then, with $\mathbf{P}^n = VD^nV^{-1}$, the probability distribution of this process at nth time step is given below:

$$\mathbf{p}^{(n)} = \mathbf{p}^{(0)}\mathbf{P}^n = \begin{bmatrix} 1 & 0 \end{bmatrix} \times \begin{bmatrix} \frac{2}{3} + \frac{1}{3}\left(\frac{1}{4}\right)^n & \frac{1}{3} - \frac{1}{3}\left(\frac{1}{4}\right)^n \\ \frac{2}{3} - \frac{2}{3}\left(\frac{1}{4}\right)^n & \frac{1}{3} + \frac{2}{3}\left(\frac{1}{4}\right)^n \end{bmatrix}$$

$$= \begin{bmatrix} \frac{2}{3} + \frac{1}{3}\left(\frac{1}{4}\right)^n & \frac{1}{3} - \frac{1}{3}\left(\frac{1}{4}\right)^n \end{bmatrix}.$$

3.2.3 Markov Process Property and Classification

In a discrete-time Markov process, we can classify a state in the state space using transition probability as follows:

- State j is said to be *accessible* from state i if $P_{ij}^{(n)} > 0$ for some $n \geq 0$. This means that if a process starts from state i, it is possible that the process will enter state j at some time n.
- Two states, state i and state j, are said to *communicate*, denoted by $i \iff j$, when they are accessible from each other, i.e., $P_{ij}^{(n)}, P_{ji}^{(m)} > 0$ for some

$n, m \geq 0$. Equivalently, if state i communicates with state j, then state j communicates with state i, and vice versa.

- Note that state i communicates with itself since we can let $n = 0$; then $P_{ii}^{(0)} = P(Z_n = i | Z_n = i) = 1 > 0$, for all $i \geq 0$.
- If state i communicates with state j and state j communicates with state k, then state i also communicates with state k. Using the Chapman-Kolmogorov equation (3.11), we can write $P_{ik}^{(n+m)} = \sum_{t=0}^{\infty} P_{it}^n P_{tk}^m \geq P_{ij}^n P_{jk}^m > 0$.

- With the communication property, a state space can be decomposed into disjoint sets, C_i, called a *communicating class*. The states are in the same communicating class when they can communicate with each other.
  - If we choose a state 0 from the state space, S, we can put state 0 and all states communicating with $0, \forall k \iff 0$, in a class $C_0 = \{0, k\}, \forall k \iff 0$. Then, we can choose a state t in the state space that is not in the class $C_0, t \in S \setminus C_0$. Since state t does not communicate with state 0 (otherwise it will be in C_0), we can put state t in a class C_1. Once all states in the state space have been assigned in classes, the state space is divided into disjoint sets, $C_i \cap C_j = \emptyset, i \neq j$ and $S = \cup_{\forall i} C_i$. Each set C_i is called a communicating class.

- A communication class C is *closed* if for any state $i \in C$; if state j does not belong to class $C, j \in S \setminus C$, then state j is not accessible from state $i, P_{ij}^{(n)} = 0$ for all $n \geq 0$. In other words, once the state is in class C, it can never escape from class C.

- If $C = \{j\}$ is closed, we call state j, an *absorbing state*, i.e., $P_{jj}^{(n)} = 1$ for all $n \geq 0$. In reliability analysis, we apply this concept to failure states where once the process enters the absorbing states, it cannot leave unless the process restarts.

- The discrete-time Markov process is said to be *irreducible* when all states communicate with each other, i.e., the state space has only one communication class. Otherwise, the process is called *reducible*.

Example 3.3 The state transition diagram of a process is shown in Figure 3.9. We can find transition probability matrix of the process in the following:

$$\mathbf{P} = \begin{pmatrix} \frac{1}{2} & \frac{1}{2} & 0 \\ \frac{1}{2} & \frac{1}{2} & 0 \\ 0 & 0 & 1 \end{pmatrix}.$$

In this example, we can find that the process has two classes, $C_0 = \{0, 1\}$ *and* $C_1 = \{2\}$, and both classes are closed. State 2 is called an absorbing state.

The previous classification concerns only the accessibility of all states in the state space. Another important classification depending on how often a state

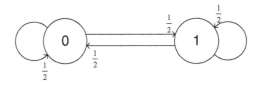

Figure 3.9 State-transition diagram of Example 3.3.

in the process can be reentered is presented in the following. This property guarantees the existence of the time taken by the process takes to reach specific states.

- State i is said to be *recurrent* if, starting from state i, the process can reach back to state i with a probability of one over a finite number of steps. This implies that once state i is recurrent, the process will reenter state i infinitely many times, i.e., $\sum_{n=1}^{\infty} P_{ii}^{(n)} = \infty$. State i is said to be *transient* if starting from state i, probability of the process reaching back to state i is less than one, i.e., there is non-zero probability of it never returning to i. This implies that if state i is transient, the process will reenter state i in a finite number of times, i.e. $\sum_{n=1}^{\infty} P_{ii}^{(n)} < \infty$.
  - If state i is recurrent and state i communicates with state j, then state j is also recurrent. This implies that if one state in a communicating class is recurrent, all states in the class are also recurrent.
  - In a finite-state discrete-time Markov process, not all states can be transient. If all states are transient states, there will be a finite time T that a process visits transient states; after that time T, no state will be visited. Since a process has to be in some state after time T, this leads us to a contradiction. Thus, at least one state has to be recurrent.
- Note that the closed classes are recurrent, and the classes that are not closed have transient states.

Example 3.4 The state transition diagram of a process is shown in Figure 3.10.

We can find the following transition probability matrix of this process as follows:

$$\mathbf{P} = \begin{pmatrix} \frac{1}{2} & \frac{1}{2} & 0 \\ \frac{1}{3} & \frac{1}{3} & \frac{1}{3} \\ 0 & 0 & 1 \end{pmatrix}.$$

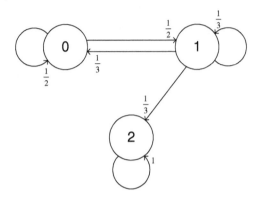

Figure 3.10 State-transition diagram of Example 3.4.

Then we can find that in this process, there are two classes, $C_0 = \{0, 1\}$ and $C_1 = \{2\}$, and class C_1 is closed. State 2 is called an absorbing state. State 0 and 1 are transient states where state 2 is a recurrent state.

In reliability analysis, we are concerned about the long-run behavior of the process. In particular, we are interested in finding the probability that the process will be in a given state, and especially in failure states. We revisit Example 3.2 to find the probability when the number of time steps approach infinity.

Example 3.5 Continuing from Example 3.2, find the probability distribution of this process at the nth time step when the initial probability is $p^{(0)} = \begin{bmatrix} p_0^{(0)} & p_1^{(0)} \end{bmatrix}$.

The probability distribution of this process at the nth time step is

$$p^{(n)} = \begin{bmatrix} \dfrac{2}{3} + \dfrac{p_0^{(0)} - 2p_1^{(0)}}{3}(\dfrac{1}{4})^n & \dfrac{1}{3} - \dfrac{p_0^{(0)} - 2p_1^{(0)}}{3}(\dfrac{1}{4})^n \end{bmatrix}.$$

Note that in Example 3.2, when the process starts from state 0, the probability distribution of this process at the nth time step is given as $\begin{bmatrix} \dfrac{2}{3} + \dfrac{1}{3}(\dfrac{1}{4})^n & \dfrac{1}{3} - \dfrac{1}{3}(\dfrac{1}{4})^n \end{bmatrix}$, and when the process starts from state 1, the probability distribution of this process at the nth time step is $p^{(n)} = \begin{bmatrix} \dfrac{2}{3} - \dfrac{2}{3}(\dfrac{1}{4})^n & \dfrac{1}{3} + \dfrac{2}{3}(\dfrac{1}{4})^n \end{bmatrix}$. If we increase the number of time steps, $n \to \infty$, then $p^{(\infty)} = \begin{bmatrix} \dfrac{2}{3} & \dfrac{1}{3} \end{bmatrix}$, which is the same no matter from where the process starts.

3.2.4 Equilibrium Distribution

From Example 3.5, it can be noticed that no matter what the initial probability is, there seems to be an *equilibrium distribution*, i.e., no matter where the

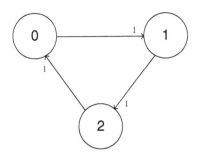

Figure 3.11 State-transition diagram of Example 3.6.

process starts from, the probability distribution at $n \to \infty$ will still be the same. The equilibrium distribution is sometimes called a stationary distribution. In order to understand the uniqueness and existence of equilibrium distribution, we need the following properties:

- State i is said to have *period d* where d is the greatest common divider of $n, n \geq 1$ when $P_{ii}^{(n)} > 0$. This means that state i maybe accessible only at times $d, 2d, 3d, \ldots$. If state i has period d and state i communicates with state j, then state j also has period d.

Example 3.6   Consider a process shown in Figure 3.11. The transition probability matrix of this process is given as follows:

$$\mathbf{P} = \begin{pmatrix} 0 & 1 & 0 \\ 0 & 0 & 1 \\ 1 & 0 & 0 \end{pmatrix}.$$

We can see that in this process, there is only one class, $C_0 = \{0, 1, 2\}$. Since this class is closed, all states are recurrent states. The transition probability at time steps 2, 3 and 4 are

$$\mathbf{P}^2 = \begin{pmatrix} 0 & 0 & 1 \\ 1 & 0 & 0 \\ 0 & 1 & 0 \end{pmatrix}.$$

$$\mathbf{P}^3 = \begin{pmatrix} 1 & 0 & 0 \\ 0 & 1 & 0 \\ 0 & 0 & 1 \end{pmatrix}.$$

$$\mathbf{P}^4 = \begin{pmatrix} 0 & 1 & 0 \\ 0 & 0 & 1 \\ 1 & 0 & 0 \end{pmatrix}.$$

This means that the process, if started from state 0 (or state 1 or 2) can return to state 0 (or state 1 or 2) in three time steps. Therefore, all states have period of three. In this process, the equilibrium distribution can only be interpreted as the proportion of time spent in each state.

- A state with period 1 is said to be *aperiodic*.
- State i is said to be *positive recurrent* if the expected time that the process takes, starting from state i, to reach back to state i is finite. In a finite-state irreducible and aperiodic process, all recurrent states are positive recurrent.
- The process is said to be *ergodic* when it is irreducible and the states are positive recurrent and aperiodic.

If the process is ergodic, $\lim_{n\to\infty} P_{ij}^{(n)}$ exists and is independent of i. The equilibrium distribution can be found in this case. If we let, $p_j^{(\infty)} = \lim_{n\to\infty} P_{ij}^{(n)}, j \geq 0$, then $p_j^{(\infty)}$ is the unique solution of (3.16) and (3.17):

$$p_j^{(\infty)} = \sum_{i=0}^{\infty} p_i^{(\infty)} P_{ij}, j \geq 0. \tag{3.16}$$

$$\sum_{j=0}^{\infty} p_j^{(\infty)} = 1. \tag{3.17}$$

Equivalently, (3.16) can be written in matrix form as $\mathbf{p}^{(\infty)} = \mathbf{p}^{(\infty)}\mathbf{P}$ together with (3.17). Intuitively, this means that if the probability is at equilibrium, it should not be changed even when the process moves in one extra time step.

Example 3.7 Continuing from Example 3.1, let us find equilibrium distribution of this process.

Since the transition probability at nth time step is

$$\mathbf{P}^n = \begin{bmatrix} \frac{2}{3} + \frac{1}{3}(\frac{1}{4})^n & \frac{1}{3} - \frac{1}{3}(\frac{1}{4})^n \\ \frac{2}{3} - \frac{2}{3}(\frac{1}{4})^n & \frac{1}{3} + \frac{2}{3}(\frac{1}{4})^n \end{bmatrix}.$$

We can find $\lim_{n\to\infty} P_{ij}^{(n)}$; then

$$\mathbf{P}^\infty = \begin{bmatrix} \frac{2}{3} & \frac{1}{3} \\ \frac{2}{3} & \frac{1}{3} \end{bmatrix},$$

which is independent of i.

The equilibrium distribution is thus $p^{(\infty)} = \begin{bmatrix} \frac{2}{3} & \frac{1}{3} \end{bmatrix}$.

We can also find equilibrium distribution from solving the following equations:

$$\begin{bmatrix} p_0^{(\infty)} & p_1^{(\infty)} \end{bmatrix} = \begin{bmatrix} p_0^{(\infty)} & p_1^{(\infty)} \end{bmatrix} \begin{bmatrix} \frac{2}{3} & \frac{1}{3} \\ \frac{2}{3} & \frac{1}{3} \end{bmatrix}.$$

Rewriting the above set of equations, we have

$$\frac{1}{3}p_0^{(\infty)} - \frac{2}{3}p_1^{(\infty)} = 0.$$

Together with $p_0^{(\infty)} + p_1^{(\infty)} = 1$, we arrive at the same solution: $p_0^{(\infty)} = \frac{2}{3}$ and $p_1^{(\infty)} = \frac{1}{3}$.

3.2.5 Mean First Passage Time

It is also of interest to find the time that a process encounters a state for the first time, called *first passage time*. In reliability analysis, we use this concept to find the time that a process enters failure states for the first time, called *time to the first failure*. Since the number of time steps to reach failure states is also probabilistic, we can then find expected value of this time, called *mean first passage time* or *mean time to first failure (MTTFF)* for reliability application.

For a finite-state discrete-time Markov process, the state space can be decomposed into disjoint sets of transient and recurrent states. For finding the mean first passage time, some transition probabilities may need to be set to zero to create appropriate sets of transient and recurrent states. Then, the state space will comprise of one transient set and multiple recurrent sets. Let $S = T \cup (\cup_{\forall i} C_i)$, where T is a transient set and C_i are closed recurrent classes, then the transition probability matrix can be rearranged as follows:

$$\mathbf{P} = \begin{bmatrix} \mathbf{Q} & \tilde{\mathbf{Q}} \\ \mathbf{0} & \tilde{\mathbf{P}} \end{bmatrix}, \tag{3.18}$$

where $\mathbf{Q} = [P_{ij}], i, j \in T$ represents transition probability between transient states, $\tilde{\mathbf{Q}} = [P_{ij}], i \in T, j \in T^c$ represents transition probability from transient states to closed recurrent class and $\tilde{\mathbf{P}} = [P_{ij}], i \in T^c, j \in T^c$ is the transition probability matrix of the states in closed recurrent classes. Note that the rest of the transition probability matrix is zero because there will be no transition from the closed recurrent class to the transient states.

Since all C_i are closed recurrent classes once the system enters a state in these classes it cannot escape back to transient states. We call states in these classes *absorbing states* when computing first passage time. We are interested in finding the time that the process spent in the transient states before absorption.

At time step 2, the transition probability matrix can be written as

$$\mathbf{P}^{(2)} = \begin{bmatrix} \mathbf{Q} & \tilde{\mathbf{Q}} \\ \mathbf{0} & \tilde{\mathbf{P}} \end{bmatrix} \times \begin{bmatrix} \mathbf{Q} & \tilde{\mathbf{Q}} \\ \mathbf{0} & \tilde{\mathbf{P}} \end{bmatrix} = \begin{bmatrix} \mathbf{Q}^2 & \mathbf{Q}\tilde{\mathbf{Q}} + \tilde{\mathbf{Q}}\tilde{\mathbf{P}} \\ \mathbf{0} & \tilde{\mathbf{P}}^2 \end{bmatrix} = \begin{bmatrix} \mathbf{Q}^{(2)} & \tilde{\mathbf{Q}}^{(2)} \\ \mathbf{0} & \tilde{\mathbf{P}}^{(2)} \end{bmatrix},$$

$$(3.19)$$

where $\mathbf{Q}^{(2)} = \mathbf{Q}\tilde{\mathbf{Q}} + \tilde{\mathbf{Q}}\tilde{\mathbf{P}}$.

At time step n, the transition probability matrix can be written as

$$\mathbf{P}^{(n)} = \begin{bmatrix} \mathbf{Q} & \tilde{\mathbf{Q}} \\ \mathbf{0} & \tilde{\mathbf{P}} \end{bmatrix}^n = \begin{bmatrix} \mathbf{Q}^{(n)} & \tilde{\mathbf{Q}}^{(n)} \\ \mathbf{0} & \tilde{\mathbf{P}}^{(n)} \end{bmatrix},$$

$$(3.20)$$

where $\mathbf{Q}^{(n)} = \mathbf{Q}^n$ and $\tilde{\mathbf{P}}^{(n)} = \tilde{\mathbf{P}}^n$.

The term $Q_{ij}^{(n)}$ gives the transition probability from transient state i to state j at time step n. Since at some point in time the process will reach absorbing states, as a consequence the summation of transition probability at all time steps has to be finite, i.e.,

$$\sum_{n=0}^{\infty} Q_{ij}^{(n)} = \sum_{n=0}^{\infty} Q_{ij}^n < \infty.$$

$$(3.21)$$

If $\sum_{n=0}^{\infty} Q_{ij}^{(n)} = \infty$, it implies that there will always be a probability that a process can move between transient states and that the process can never reach absorbing states. This contradicts the fact that the time spent in the transient states is finite.

Let N_{ij} be an expected number of time steps that the process takes from transient state i to state j just before reaching absorbing states; then, using conditional probability we have

$$N_{ii} = 1 + \sum_{k=1}^{|T|} Q_{ik} N_{ki}.$$

$$(3.22)$$

Equation (3.22) yields the expected number of time steps from going from state i to itself. The first term "1" represents the time steps needed to take the process out from state i, and the latter term represents the time steps from other transient states to state i based on conditional probability.

$$N_{ij} = 0 + \sum_{k=1}^{|T|} Q_{ik} N_{kj}$$

$$(3.23)$$

Equation (3.23) yields the expected number of time steps from transient state i to state j using conditional probability. We can write (3.22) and (3.23) in the matrix form as follows:

$$\mathbf{N} = \mathbf{I} + \mathbf{Q}\mathbf{N},$$

$$(3.24)$$

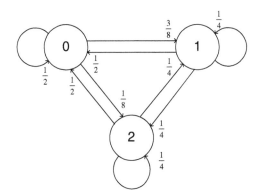

Figure 3.12 State-transition diagram in Example 3.8.

where $\mathbf{I}$ is the identity matrix of size $|T|$. The expected number of time steps, $\mathbf{N}$, can be found from

$$\mathbf{N} = (\mathbf{I} - \mathbf{Q})^{-1}. \tag{3.25}$$

The matrix $\mathbf{N}$ is called *fundamental matrix*. Since an element N_{ij} yields the expected time steps from transient state i to state j just before reaching absorbing states, we can find the expected time steps, n_i, to reach an absorbing state from state i as follows:

$$n_i = \sum_{k=1}^{|T|} N_{ik}. \tag{3.26}$$

The example of mean first passage time is given below.

Example 3.8 A process can be represented by the state-transition diagram shown in Figure 3.12. Find the mean first passage time to state 1, starting from state 0 or state 2.

To solve this problem, state 1 is the absorbing state. With state 1 as absorbing state, the transition probability matrix between transient states is $\mathbf{Q}^{(n)} =$
$\begin{bmatrix} \frac{1}{2} & \frac{1}{8} \\ \frac{1}{2} & \frac{1}{4} \end{bmatrix}$. The fundamental matrix is found from (3.25):

$$\mathbf{N} = \begin{bmatrix} 1 - \frac{1}{2} & -\frac{1}{8} \\ -\frac{1}{2} & 1 - \frac{1}{4} \end{bmatrix}^{-1} = \begin{bmatrix} \frac{12}{5} & \frac{2}{5} \\ \frac{8}{5} & \frac{8}{5} \end{bmatrix}.$$

This matrix gives the number of time steps in transient states before reaching absorbing state 1. For example, $N_{02} = \frac{2}{5}$ is the expected number of time steps from state 0 to state 2 before hitting the absorbing state 1.

The expected first passage time if the system starts from state 0 is $n_0 = N_{00} + N_{02} = \frac{14}{5}$, and the expected first passage time if the system starts from state 2 is $n_2 = N_{20} + N_{22} = \frac{16}{5}$.

Alternatively, the fundamental matrix N can be derived by finding the number of time steps that the absorbing states can be reached in the 1st, 2nd, 3rd, ..., and nth time step as follows:

$$N = I + Q + Q^2 + Q^3 + \dots \tag{3.27}$$

Rearranging equation (3.27), we have

$$N = I + Q(I + Q + Q^2 + Q^3 + \dots) = I + QN. \tag{3.28}$$

The solution to this equation is the same as (3.25).

3.3 Continuous-Time Markov Process

The continuous-time Markov process is the stochastic process with continuous index set $\{Z_t, t \geq 0\}$. Recall from the Markov property (3.29):

$$P(Z_{s+t} = j | Z_s = i, Z_u = a_u, \forall u \in [0, s)) = P(Z_{s+t} = j | Z_s = i). \tag{3.29}$$

This property implies that the probability of transition from state i to state j does not depend on the initial start time, s, but depends on the elapsed time, t. Thus, the transition probability from state i to state j in terms of this time t, denoted by $P_{ij}(t)$, can be written as follows:

$$P_{ij}(t) = P(Z_{s+t} = j | Z_s = i). \tag{3.30}$$

It follows that $0 \leq P_{ij}(t) \leq 1, \forall t$ and $\sum_{j=0}^{\infty} P_{ij}(t) = 1, \forall t, i = 0, 1, 2, \dots$

For a stationary continuous-time Markov process, we have the following:

$$P(Z_{s+t} = j | Z_s = i) = P(Z_t = j | Z_0 = i). \tag{3.31}$$

When a process is moving from state i to state j in time $t + s$, we can write a Chapman-Kolmogorov equation for a continuous time process as (3.33).

$$P_{ij}(t + s) = P(Z_{t+s} = j | Z_0 = i) \tag{3.32}$$

$$= \sum_{k=0}^{\infty} P(Z_{t+s} = j, Z_t = k | Z_0 = i)$$

$$= \sum_{k=0}^{\infty} P(Z_{t+s} = j | Z_t = k, Z_0 = i) \times P(Z_t = k | Z_0 = i)$$

$$= \sum_{k=0}^{\infty} P(Z_{t+s} = j | Z_t = k) \times P(Z_t = k | Z_0 = i)$$

$$= \sum_{k=0}^{\infty} P_{kj}(s) P_{ik}(t).$$

The transition probability of a continuous-time process is time dependent; let X_{ij} denote the time that the process spends to transit from state i to state j, then the transition probability from state i to state j can be given by (3.33):

$$P_{ij}(t) = P(s < X_{ij} < s + t | X_{ij} > s). \tag{3.33}$$

The transition probability (3.33) is the probability that a process will move to state j within elapsed time t given that the process has been in state i for some time s.

Define a transition rate from state i to state j

$$\lambda_{ij} = \frac{dP_{ij}(t)}{dt} \tag{3.34}$$

Since the process is stationary, the transition rate is constant and independent of time t. Note that in Chapter 2, the only random variable that has a memoryless property with a constant hazard function is the exponential random variable. This means that the time between transition of the states in the continuous-time Markov process is exponentially distributed.

As $\Delta t \to 0$,

$$\lambda_{ij} = \lim_{\Delta t \to 0} \frac{P_{ij}(\Delta t) - P_{ij}(0)}{\Delta t}. \tag{3.35}$$

Then, we can write

$$\lambda_{ij}\Delta t = P_{ij}(\Delta t) - P_{ij}(0). \tag{3.36}$$

When $i = j$,

$$P_{ij}(0) = 1, P_{ii}(\Delta t) = 1 + \lambda_{ii}\Delta t,$$

and when $i \neq j$,

$$P_{ij}(0) = 0, P_{ij}(\Delta t) = \lambda_{ij}\Delta t.$$

Note that $\sum_j P_{ij}(\Delta t) = 1$; we have

$$1 + \lambda_{ii}\Delta + \sum_{j, j \neq i} \lambda_{ij}\Delta t = 1. \tag{3.37}$$

This gives (3.38):

$$\lambda_{ii} = -\sum_{j, j \neq i} \lambda_{ij}. \tag{3.38}$$

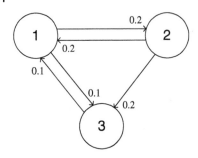

Figure 3.13 State-transition diagram of a three-state system continuous-time process.

3.3.1 Transition Rate Matrix

We can write the transition rate between each state in the state space in a matrix form, called **R**, the *transition rate matrix*, as follows:

$$\mathbf{R} = [\lambda_{ij}] = \begin{bmatrix} \lambda_{00} & \lambda_{01} & \cdots \\ \lambda_{10} & \lambda_{11} & \cdots \\ \vdots & \vdots & \vdots \end{bmatrix}. \tag{3.39}$$

This matrix gives the transition rate from one state to others. Since the process is stationary, this transition rate matrix is constant. Similar to the discrete-time case, we can represent the process by a state-transition diagram that shows all possible states as well as the transition rates. Figure 3.13 shows the state-transition diagram of a three-state system as an example.

Example 3.9 Write the transition rate matrix of the system in Figure 3.13:

$$\mathbf{R} = [\lambda_{ij}] = \begin{bmatrix} -0.3 & 0.2 & 0.1 \\ 0.2 & -0.4 & 0.2 \\ 0.1 & 0 & -0.1 \end{bmatrix}.$$

Using (3.33) to $P_{ij}(t + \Delta t)$, we have

$$P_{ij}(t + \Delta t) = \sum_{k} P_{ik}(t) P_{kj}(\Delta t) = P_{ij}(t) P_{jj}(\Delta t) + \sum_{k, k \neq j} P_{ik}(t) P_{kj}(\Delta t).$$

Substitute with (3.36), and we have

$$P_{ij}(t + \Delta t) = P_{ij}(t)(1 + \lambda_{jj}\Delta t) + \sum_{k, k \neq j} P_{ik}(t) \lambda_{kj}\Delta t = P_{ij}(t) + \sum_{k} P_{ik}(t) \lambda_{kj}\Delta t.$$

Then,

$$P'_{ij}(t) \equiv \frac{P_{ij}(t + \Delta t) - P_{ij}(t)}{\Delta t} = \sum_{k} P_{ik}(t) \lambda_{kj}. \tag{3.40}$$

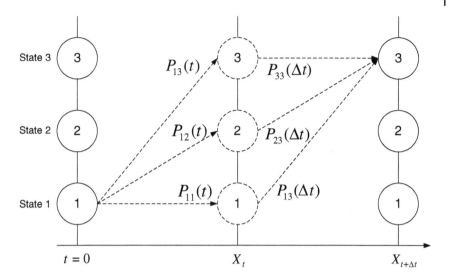

Figure 3.14 A transition from state 1 to state 3 of a continuous-time three-state system.

We can write the transition probability at time t in matrix form as follows:

$$\mathbf{P}'(t) = \mathbf{P}(t)\mathbf{R}, \tag{3.41}$$

where $\mathbf{P}(t) = [P_{ij}(t)]$ is the transition probability matrix at time t and $\mathbf{R} = [\lambda_{ij}]$ is the transition rate matrix.

We now illustrate the concept of transition rate by a three-state system shown in Figure 3.14.

By using conditional probability concept as $\Delta t \to 0$,

$$P_{13}(t + \Delta t) = P_{11}(t)P_{13}(\Delta t) + P_{12}(t)P_{23}(\Delta t) + P_{13}(t)P_{33}(\Delta t).$$

Using (3.36), we can write

$$P_{13}(t + \Delta t) = P_{11}(t)\lambda_{13}\Delta t + P_{12}(t)\lambda_{23}\Delta t + P_{13}(t)(1 + \lambda_{33}\Delta t).$$

Then,

$$P_{13}(t + \Delta t) - P_{13}(t) = P_{11}(t)\lambda_{13}\Delta t + P_{12}(t)\lambda_{23}\Delta t + P_{13}(t)\lambda_{33}\Delta t$$

$$\frac{P_{13}(t + \Delta t) - P_{13}(t)}{\Delta t} = P_{11}(t)\lambda_{13} + P_{12}(t)\lambda_{23} + P_{13}(t)\lambda_{33}$$

$$= P'_{13}(t).$$

Rewrite in the matrix form, we have

$$P'_{13}(t) = \begin{bmatrix} P_{11}(t) & P_{12}(t) & P_{13}(t) \end{bmatrix} \begin{bmatrix} \lambda_{13} \\ \lambda_{23} \\ \lambda_{33} \end{bmatrix}.$$

Since $P_{13}(\Delta t) + P_{23}(\Delta t) + P_{33}(\Delta t) = 1$, then $\lambda_{33} = -\lambda_{13} - \lambda_{23}$. Similarly, we can write

$$\begin{bmatrix} P'_{11}(t) & P'_{12}(t) & P'_{13}(t) \end{bmatrix} = \begin{bmatrix} P_{11}(t) & P_{12}(t) & P_{13}(t) \end{bmatrix} \mathbf{R},$$

where $\mathbf{R} = \begin{bmatrix} \lambda_{12} & \lambda_{12} & \lambda_{13} \\ \lambda_{21} & \lambda_{22} & \lambda_{23} \\ \lambda_{31} & \lambda_{32} & \lambda_{33} \end{bmatrix}$ is the transition rate matrix for a three-state system.

We can find transition probability matrix at time t by solving the first order equation (3.41). Since $\mathbf{P}(0) = \mathbf{I}$, the solution is given in (3.42):

$$\mathbf{P}(t) = e^{\mathbf{R}t}, \tag{3.42}$$

where $e^{\mathbf{R}t}$ can be found using matrix decomposition, $e^{\mathbf{R}t} = Ve^{Dt}V^{-1}$; D is a diagonal matrix of eigen values, and columns of V are corresponding eigen vectors. We can use this decomposition to find time-specific distribution function of the process in a closed form.

3.3.2 Probability Distribution at Time t

Equation (3.42) gives the transition probability at time t; using the transition probability, we can find the distribution of the process at time t, given that the initial distribution at time $0, Z_0$, is known. Define the probability distribution at initial time by (3.43):

$$\mathbf{p}(0) = [p_0(0), p_1(0), \dots, p_k(0), \dots]. \tag{3.43}$$

Note that $p_k(0)$ is the probability that the process starts from state k at time 0, i.e., $p_k(0) = P(Z_0 = k)$, and $\sum_{k=0}^{\infty} p_k(0) = 1$.

The *probability distribution at time t* can be found using (3.44):

$$\mathbf{p}(t) = \mathbf{p}(0)\mathbf{P}(t). \tag{3.44}$$

Recall the derivative of transition probability from (3.40); we can rearrange the equation in a matrix form as follows:

$$\mathbf{P}(t + \Delta t) = \mathbf{P}(t) + \mathbf{P}'(t)\Delta t. \tag{3.45}$$

Substitute $\mathbf{P}'(t) = \mathbf{P}(t)\mathbf{R}$ from (3.41), and we have

$$\mathbf{P}(t + \Delta t) = \mathbf{P}(t)(\mathbf{I} + \mathbf{R}\Delta t). \tag{3.46}$$

Note that (3.46) yields a transition probability at the time $t + \Delta t$, which is found from the multiplication of transition probability at time t and $\mathbf{I} + \mathbf{R}\Delta t$. Rearrange (3.46). Let

$$\mathbf{P}(\Delta t) = [\mathbf{I} + \mathbf{R}\Delta t]. \tag{3.47}$$

Using both (3.46) and (3.47), we can write

$$\mathbf{P}(t + \Delta t) = \mathbf{P}(t)\mathbf{P}(\Delta t). \tag{3.48}$$

This means that we can approximate the transition probability matrix using (3.47); the transition probability at time step j or equivalently at time $j\Delta t$ is given by (3.49).

$$\mathbf{P}(j\Delta t) = [\mathbf{I} + \mathbf{R}\Delta t]^j. \tag{3.49}$$

We approximate the continuous-time process by the discrete one, and the probability distribution at any time step can now be calculated from (3.15).

Example 3.10   Consider a system with two states shown in Figure 3.15. Find the transition probability matrix and probability distribution at time t if the system starts from state 0 and 1, correspondingly.
The transition rate matrix of this process is

$$\mathbf{R} = \begin{bmatrix} -\lambda & \lambda \\ \mu & -\mu \end{bmatrix}.$$

The transition probability matrix can be found from (3.42); we first calculate the eigen value and eigen vector from $det(\mathbf{R} - d\mathbf{I}) = 0$, and we have $d = 0, -(\lambda + \mu)$.
For eigen value of 0, the eigen vector can be found from

$$\begin{bmatrix} -\lambda & \lambda \\ \mu & -\mu \end{bmatrix} \begin{bmatrix} v_{11} \\ v_{21} \end{bmatrix} = \begin{bmatrix} 0 \\ 0 \end{bmatrix}$$

Figure 3.15 A two-state component model.

Then $v_{11} = v_{21}$; let $v_{11} = 1$, $\begin{bmatrix} v_{11} \\ v_{21} \end{bmatrix} = \begin{bmatrix} 1 \\ 1 \end{bmatrix}$. The eigen vector for $d = -(\lambda + \mu)$,

$$\begin{bmatrix} \mu & \lambda \\ \mu & \lambda \end{bmatrix} \begin{bmatrix} v_{12} \\ v_{22} \end{bmatrix} = \begin{bmatrix} 0 \\ 0 \end{bmatrix}$$

Then $v_{12} = -\frac{\lambda}{\mu} v_{22}$; let $v_{22} = \mu$, $\begin{bmatrix} v_{12} \\ v_{22} \end{bmatrix} = \begin{bmatrix} -\lambda \\ \mu \end{bmatrix}$.

Then,

$$\mathbf{P}(t) = e^{\mathbf{R}t}$$

$$= \begin{bmatrix} 1 & -\lambda \\ 1 & \mu \end{bmatrix} \times \begin{bmatrix} e^{0t} & 0 \\ 0 & e^{-(\lambda+\mu)t} \end{bmatrix} \times \begin{bmatrix} 1 & -\lambda \\ 1 & \mu \end{bmatrix}^{-1}$$

$$= \begin{bmatrix} \frac{\mu}{\lambda+\mu} + \frac{\lambda}{\lambda+\mu}e^{-(\lambda+\mu)t} & \frac{\lambda}{\lambda+\mu} - \frac{\lambda}{\lambda+\mu}e^{-(\lambda+\mu)t} \\ \frac{\mu}{\lambda+\mu} - \frac{\mu}{\lambda+\mu}e^{-(\lambda+\mu)t} & \frac{\lambda}{\lambda+\mu} + \frac{\mu}{\lambda+\mu}e^{-(\lambda+\mu)t} \end{bmatrix}.$$

The probability distribution of this process at time t if the system starts from state 0 is

$$\mathbf{p}(t) = \mathbf{p}(0)\mathbf{P}(t)$$

$$= \begin{bmatrix} 1 & 0 \end{bmatrix} \times \begin{bmatrix} \frac{\mu}{\lambda+\mu} + \frac{\lambda}{\lambda+\mu}e^{-(\lambda+\mu)t} & \frac{\lambda}{\lambda+\mu} - \frac{\lambda}{\lambda+\mu}e^{-(\lambda+\mu)t} \\ \frac{\mu}{\lambda+\mu} - \frac{\mu}{\lambda+\mu}e^{-(\lambda+\mu)t} & \frac{\lambda}{\lambda+\mu} + \frac{\mu}{\lambda+\mu}e^{-(\lambda+\mu)t} \end{bmatrix}$$

$$= \begin{bmatrix} \frac{\mu}{\lambda+\mu} + \frac{\lambda}{\lambda+\mu}e^{-(\lambda+\mu)t} & \frac{\lambda}{\lambda+\mu} - \frac{\lambda}{\lambda+\mu}e^{-(\lambda+\mu)t} \end{bmatrix}.$$

The probability distribution of this process at time t if the system starts from state 1 is

$$\mathbf{p}(t) = \mathbf{p}(0)\mathbf{P}(t)$$

$$= \begin{bmatrix} 0 & 1 \end{bmatrix} \times \begin{bmatrix} \frac{\mu}{\lambda+\mu} + \frac{\lambda}{\lambda+\mu}e^{-(\lambda+\mu)t} & \frac{\lambda}{\lambda+\mu} - \frac{\lambda}{\lambda+\mu}e^{-(\lambda+\mu)t} \\ \frac{\mu}{\lambda+\mu} - \frac{\mu}{\lambda+\mu}e^{-(\lambda+\mu)t} & \frac{\lambda}{\lambda+\mu} + \frac{\mu}{\lambda+\mu}e^{-(\lambda+\mu)t} \end{bmatrix}$$

$$= \begin{bmatrix} \frac{\mu}{\lambda+\mu} - \frac{\mu}{\lambda+\mu}e^{-(\lambda+\mu)t} & \frac{\lambda}{\lambda+\mu} + \frac{\mu}{\lambda+\mu}e^{-(\lambda+\mu)t} \end{bmatrix}.$$

Note that the probability distribution at time $t \to \infty$ is the same whether the process starts from state 0 or 1.

3.3.3 Equilibrium Distribution

Similar to the discrete-time case if the process is ergodic, $\lim_{t\to\infty} P_{ij}(t)$ exists and is independent of i. Let $p_j^{(\infty)} = \lim_{t\to\infty} P_{ij}(t), j \geq 0; p_j^{(\infty)}$ is the equilibrium distribution for a continuous-time process.

From (3.40), $P'_{ij}(t) = \sum_k P_{ik}(t)\lambda_{kj}$; take the limit and let $t \to \infty$, assume that the limit and summation can be interchanged, and we have

$$\lim_{t\to\infty} P'_{ij}(t) = \lim_{t\to\infty} \sum_k P_{ik}(t)\lambda_{kj} \tag{3.50}$$

$$= \sum \lim_{t\to\infty} P_{ik}(t)\lambda_{kj} \tag{3.51}$$

$$= \sum p_j^{(\infty)}\lambda_{kj}. \tag{3.52}$$

Since $P'_{ij}(t)$ is the rate of transition probability matrix, which at a steady state, $t \to \infty$, will converge to zero. Then,

$$0 = \sum p_j^{(\infty)}\lambda_{kj}. \tag{3.53}$$

Rewrite (3.53) in a matrix form, and we have

$$\mathbf{p}^{(\infty)}\mathbf{R} = \mathbf{0} \tag{3.54}$$

Together with $\sum_{j=0}^{\infty} p_j^{(\infty)} = 1$, we can find the equilibrium distribution of the process.

Example 3.11 Let us find the equilibrium distribution of the process in Example 3.10. From the solution in Example 3.10, we can find $p_j^{(\infty)} = \lim_{t\to\infty} P_{ij}(t)$; then,

$$\mathbf{P}^{\infty} = \begin{bmatrix} \frac{\mu}{\lambda+\mu} & \frac{\lambda}{\lambda+\mu} \\ \frac{\mu}{\lambda+\mu} & \frac{\lambda}{\lambda+\mu} \end{bmatrix},$$

which is independent of i.

The equilibrium distribution is equal to

$$\mathbf{p}^{(\infty)} = \begin{bmatrix} \frac{\mu}{\lambda+\mu} & \frac{\lambda}{\lambda+\mu} \end{bmatrix}.$$

We can also find equilibrium distribution from solving the following equations:

$$\begin{bmatrix} 0 & 0 \end{bmatrix} = \begin{bmatrix} p_0^{(\infty)} & p_1^{(\infty)} \end{bmatrix} \begin{bmatrix} -\lambda & \lambda \\ \mu & -\mu \end{bmatrix}.$$

Together with $p_0^{(\infty)} + p_1^{(\infty)} = 1$, we arrive at the same solution, $p_0^{(\infty)} = \frac{\mu}{\lambda+\mu}$ and $p_1^{(\infty)} = \frac{\lambda}{\lambda+\mu}$.

3.3.4 Mean First Passage Time

For a continuous-time process, we are also interested to find the time that a process enters failure states for the first time, or first passage time. Since the time to reach failure states is also probabilistic, we can then find expected value of this time, called mean first passage time or mean time to first failure (MTTFF), for reliability application.

We can compute this mean first passage time using equation (3.25) in the discrete-time process. Since the continuous-time process can be approximated by a discrete-time case where time is advanced by Δt, the transition probability matrix is approximated by (3.47), $\mathbf{P}(\Delta t) = [\mathbf{I} + \mathbf{R}\Delta t]$. The mean first passage time is given in (3.55):

$$\bar{\mathbf{T}} = \Delta t \times \mathbf{N} = \Delta t (\mathbf{I} - (\mathbf{I} + \mathbf{R}\Delta t))^{-1} = -R^{-1}. \tag{3.55}$$

Exercises

3.1 A system consists of two components. The failure rate of each component is 10 failures per year. The mean time to repair of each component is six hours. When both components fail, the system is considered failed. Assume that both components are independent and are working at the beginning of the process. (i) Draw the state transition diagram for this system. (ii) Write the transition-rate matrix of this system. (iii) Find the exact transition probability at time t. (iv) Find a failure probability at time t. (v) Find the failure probability at each hour for four hours using transition probability approximation, $\delta t = \frac{1}{4}$ hour. (vi) Find equilibrium state probability.

3.2 In order to model the deterioration state of machine insulation, a continuous-time Markov model is developed as shown in Figure 3.16. D1 represents a normal state, D2 represents deterioration stage and F denotes insulation failure state due to deterioration, where $\lambda_{12} = \frac{1}{100}$ per day, $\lambda_{2f} = \frac{1}{200}$ per day, $\mu_{21} = \frac{1}{2}$ per day and $\mu = \frac{1}{4}$ per day. If the system starts from D1, find the failure probability after four hours using the transition probability approximation with $\delta t = 2$ hours. What is the average time that the process takes from state D1, and for D2 to reach the failure state for the first time?

Figure 3.16 Machine insulation deterioration states.

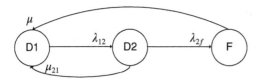

4

Frequency-Based Approach to Stochastic Process

4.1 Introduction

In the previous chapter we modeled the stochastic behavior of a system using Markov processes. When the process is ergodic, we can find the steady state probability of each state in a Markov process as well as the mean first passage time. These two indices are of importance for power system reliability analysis because they provide information on system probability of failure and the average time it takes the system to reach the failure states for the first time. However, both the steady state probability and the mean time to first failure lack some important information: *how often* the system experiences failure states. Consider the two systems shown in Figure 4.1.

It can be seen that system A is the same as in Examples 3.10 and 3.11, where the steady state probabilities of systems A and B being in the down state are given as (4.1) and (4.2):

$$p_A = \frac{\lambda}{\lambda + \mu} \tag{4.1}$$

$$p_B = \frac{4\lambda}{4\lambda + 4\mu} = \frac{\lambda}{\lambda + \mu}. \tag{4.2}$$

This means that both systems have the same probability of failure. However, the main difference between the two systems is that for system B, the transitions in both directions are four times higher than those in system A. This difference is the major concern for the systems economics and operation, as more frequent failures or repairs cause more interruptions and resources required to fix them. If the system operator uses only the probability index to evaluate system reliability, this important factor of how frequently the system fails will be missed. This means that the probability index does not reflect the differences in the failure and repair rates of both systems, which causes a major effect in system economics and operation.

Electric Power Grid Reliability Evaluation: Models and Methods, First Edition. Chanan Singh,
Panida Jirutitijaroen, and Joydeep Mitra.
© 2019 by The Institute of Electrical and Electronic Engineers, Inc. Published 2019 by John Wiley & Sons, Inc.

Figure 4.1 Two systems: System A compared to System B.

4.2 Concept of Transition Rate

Before we give the formal definition of frequency, we revisit the concept of transition rate, whose definition is given by (3.34). In this chapter, we consider the concept of transition rate in a practical perspective. The definition of transition rate is used to describe the behavior of a system in moving from one state to other states.

Consider a system with state space S. The transition rate from state i to state j is the average number of transitions from state i to state j per unit of the time spent in state i where $i, j \in S$. If the system is observed for T hours, and during this period of observation T_i hours are spent in state i, then the transition rate from state i to state j is given by (4.3). The unit of the transition rate is transitions per unit time.

$$\lambda_{ij} = \lim_{T_i \to \infty} \frac{n_{ij}}{T_i}, \tag{4.3}$$

where n_{ij} = number of transitions from state i to state j during the period of observation.

Example 4.1 Consider a system consisting of two states, namely Up and Down states shown in Figure 4.2

Let the Up state be state 1 and Down state be state 2. When we observe the system for time T, we find that the system spends the total amount of time T_1 in the up state and total amount of time T_2 in the down state. We also observe that the number of transitions from up to down state is n_{12}, where the number of transitions from down to up state is n_{21}. We can then find the transition rates from up to down state from (4.4) and the transition rate from down to up state from (4.5):

$$\lambda_{12} = \frac{n_{12}}{T_1} \tag{4.4}$$

$$\lambda_{21} = \frac{n_{21}}{T_2}. \tag{4.5}$$

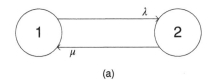

(a)

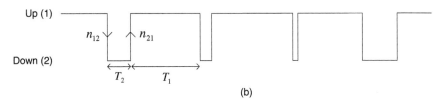

(b)

Figure 4.2 A two-state component model.

It should be noted that the average time the system spends in the Up state, denoted by *Mean Up Time* or *MUT*, can be found from (4.6):

$$MUT = \frac{T_1}{n_{12}}, \tag{4.6}$$

which is a reciprocal of the transition rate λ_{12}.

Similarly, we can find the average time the system spent in the Down state, denoted by *Mean Down Time* or *MDT*, from (4.7):

$$MDT = \frac{T_2}{n_{21}},$$

which is a reciprocal of the transition rate λ_{21}.

Typically for a system with two states, we call the transition from success to failure states as *failure rate*, λ, and the transition from failure to success states *repair rate*, μ.

The concept of transition rate has a very close relationship with the concept of frequency, in that they both represent the transition of states in the system. We introduce the concept of frequency and derive the relationship between transition rate and frequency in the next section.

4.3 Concept of Frequency

The frequency of encountering a transition to state j from state i, denoted by $Fr_{\{i\} \to \{j\}}$, is defined as the expected or average number of transitions from state i to state j per unit time, i.e.,

$$Fr_{\{i\} \to \{j\}} = \lim_{T \to \infty} \frac{n_{ij}}{T}. \tag{4.7}$$

Note that we can introduce the term T_i, or the time spent in state i, as follows:

$$Fr_{\{i\}\to\{j\}} = \lim_{T\to\infty} \frac{T_i}{T} \cdot \frac{n_{ij}}{T_i}. \tag{4.8}$$

Since the long-run fraction of time spent in state i can be described using the probability of being in state i, denoted by p_i, equation (4.8) becomes (4.9):

$$Fr_{\{i\}\to\{j\}} = p_i\lambda_{ij}. \tag{4.9}$$

Therefore, the steady state frequency of transition from state i to state j can be obtained simply by multiplying the steady state probability of state i with the transition rate from state i to state j.

This simple interpretation of frequency is very helpful in writing state equations using the frequency balance concept in Section 4.4, and the equivalent transition rate between subsets of states in Section 4.5 allowing reduction of state space. These ideas of transition rate and transition frequency can be illustrated by taking an example of a two-state component.

Example 4.2 For the same system in Example 4.1, the transition rate from Up to Down state, i.e., failure rate, λ, is given by (4.4), and the transition rate from Down to Up state, i.e., repair rate, μ, is given by (4.5). We can find the frequencies as follows:

$$Fr_{\{1\}\to\{2\}} = p_1\lambda. \tag{4.10}$$

Substitute the probability of being in state 1 from Example 3.11, and we have

$$Fr_{\{1\}\to\{2\}} = \frac{\mu}{\lambda + \mu} \times \lambda. \tag{4.11}$$

Similarly,

$$Fr_{\{2\}\to\{1\}} = \frac{\lambda}{\lambda + \mu} \times \mu. \tag{4.12}$$

It can be observed that $Fr_{\{1\}\to\{2\}} = Fr_{\{2\}\to\{1\}}$, which indicates a frequency balance between the two states of this system. This is not a coincidence and will be investigated in Section 4.4.

The concept of frequency between two states can be extended to the frequency between two disjoint sets, as explained in the following section.

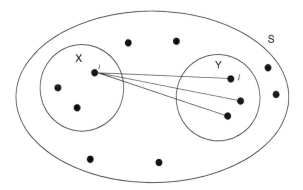

Figure 4.3 Frequency of encountering subset Y from subset X.

4.3.1 Frequency between Two Disjoint Sets

We can also find *frequency between two disjoint sets*. Intuitively, the frequency between two sets can be found by drawing a boundary around the two sets and finding the expected transition rate into the boundary or out of the boundary. Figure 4.3 shows subsets $X, Y \subset S, X \cap Y = \emptyset$. The frequency of a subset Y from a subset X is simply the addition of inter set frequencies as follows:

$$Fr_{X \to Y} = \sum_{i \in X} \left(p_i \sum_{j \in Y} \lambda_{ij} \right). \tag{4.13}$$

Note that if $Y = A \cup B, A \cap B = \emptyset$, it also follows from (4.13) that

$$Fr_{X \to (A \cup B)} = Fr_{X \to A} + Fr_{X \to B}. \tag{4.14}$$

We illustrate this concept in Figure 4.4.

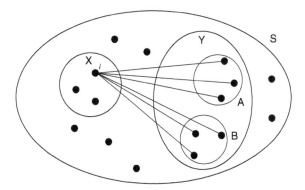

Figure 4.4 Frequency of encountering subset Y from subset X when $Y = A \cup B$ and $A \cap B = \emptyset$.

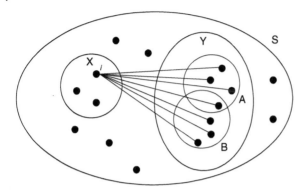

Figure 4.5 Frequency of encountering subset Y from subset X when $Y = A \cup B$ but $A \cap B \neq \emptyset$.

However, if $Y = A \cup B$ but $A \cap B \neq \emptyset$, we can write $A \cup B = (A \setminus B) \cup (B \setminus A) \cup (A \cap B)$. Since $(A \setminus B) \cap (B \setminus A) = \emptyset, (A \setminus B) \cap (A \cap B) = \emptyset$ and $(B \setminus A) \cap (A \cap B) = \emptyset$, it follows from (4.14) that

$$
\begin{aligned}
Fr_{X \to (A \cup B)} &= Fr_{X \to A \setminus B} + Fr_{X \to B \setminus A} + Fr_{X \to A \cap B} \qquad (4.15) \\
&= (Fr_{X \to A \setminus B} + Fr_{X \to A \cap B}) + (Fr_{X \to B \setminus A} + Fr_{X \to A \cap B}) - Fr_{X \to A \cap B} \\
&= Fr_{X \to A} + Fr_{X \to B} - Fr_{X \to A \cap B}.
\end{aligned}
$$

We illustrate this concept in Figure 4.5. It should also be observed that the expression above is the same concept as how we calculate the probability of the union of two events.

In general, we can write $Y = E_1 \cup E_2 \cup \ldots \cup E_n$ and write a more general expression with (4.16):

$$
\begin{aligned}
Fr_{X \to (E_1 \cup E_2 \cup \ldots \cup E_n)} &= \sum_i Fr_{X \to E_i} - \sum_{i<j} Fr_{X \to E_i \cap E_j} \qquad (4.16) \\
&+ \sum_{i<j<k} Fr_{X \to E_i \cap E_j \cap E_k} - \cdots \\
&+ (-1)^{n-1} Fr_{X \to E_1 \cap E_2 \cap \ldots \cap E_n}.
\end{aligned}
$$

When X is the set of success events and Y is the set of failure events, then the frequency from X to Y is the frequency from success to failure events. In Chapter 5 we will use (4.16) to help us calculate this frequency.

In the following section, we focus the analysis to the *frequency of a set*, i.e., the frequency of entering a set from the rest of the state space.

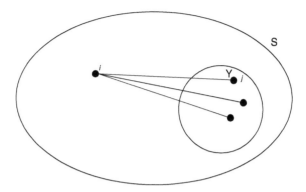

Figure 4.6 Frequency of encountering subset Y.

4.3.2 Frequency of a Set

The frequency of encountering a set, Y, from the rest of the state space shown in Figure 4.6 can be found from (4.17). We denote this frequency as $Fr_{S\setminus Y\to Y}$, or shortened as, $Fr_{\to Y}$.

$$Fr_{\to Y} = \sum_{i\in S\setminus Y} \left(p_i \sum_{j\in Y} \lambda_{ij} \right) \tag{4.17}$$

Similarly, the frequency of exiting a set, Y, to the rest of the state space, denoted by $Fr_{Y\to}$, can be found from (4.18):

$$Fr_{Y\to} = \sum_{i\in Y} \left(p_i \sum_{j\in S\setminus Y} \lambda_{ij} \right). \tag{4.18}$$

The main application in reliability analysis of both (4.17) and (4.18) is to find the frequency of entering or leaving failure states.

From (4.17) and (4.13), since $S = X \cup \bar{X}$ and $X \cap Y = \emptyset$, we can write

$$Fr_{\to Y} = \sum_{i\in X} \left(p_i \sum_{j\in Y} \lambda_{ij} \right) + \sum_{i\in \bar{X}\setminus Y} \left(p_i \sum_{j\in Y} \lambda_{ij} \right) \tag{4.19}$$

$$= Fr_{X\to Y} + Fr_{\bar{X}\setminus Y\to Y}. \tag{4.20}$$

Similarly,

$$Fr_{Y\to} = \sum_{i\in Y} \left(p_i \sum_{j\in X} \lambda_{ij} \right) + \sum_{i\in Y} \left(p_i \sum_{j\in \bar{X}\setminus Y} \lambda_{ij} \right) \tag{4.21}$$

$$= Fr_{Y\to X} + Fr_{Y\to \bar{X}\setminus Y}. \tag{4.22}$$

4.3.3 Frequency of a Union of Disjoint Sets

For disjoint events, we can derive an expression for frequency of entering *a union of disjoint events* as follows:

$$Fr_{\to X \cup Y} = \sum_{i \in S \setminus (X \cup Y)} \left(p_i \sum_{j \in X \cup Y} \lambda_{ij} \right) \tag{4.23}$$

$$= \sum_{i \in S \setminus (X \cup Y)} p_i \left(\sum_{j \in X} \lambda_{ij} + \sum_{j \in Y} \lambda_{ij} \right)$$

$$= \sum_{i \in S \setminus (X \cup Y)} p_i \left(\sum_{j \in X} \lambda_{ij} \right) + \sum_{i \in S \setminus (X \cup Y)} p_i \left(\sum_{j \in Y} \lambda_{ij} \right).$$

When $X \cap Y = \emptyset$, $S \setminus (X \cup Y) = (S \setminus X) \cap (S \setminus Y) = \bar{X} \setminus Y = \bar{Y} \setminus X$, we can rewrite (4.23) as (4.24):

$$Fr_{\to X \cup Y} = Fr_{\bar{Y} \setminus X \to X} + Fr_{\bar{X} \setminus Y \to Y}. \tag{4.24}$$

Substitute $Fr_{\bar{X} \setminus Y \to Y} = Fr_{\to Y} - Fr_{X \to Y}$ from (4.19), and we have

$$Fr_{\to X \cup Y} = Fr_{\to X} - Fr_{Y \to X} + Fr_{\to Y} - Fr_{X \to Y}. \tag{4.25}$$

Similarly, the frequency of exiting $X \cup Y$ can be found from (4.26):

$$Fr_{X \cup Y \to} = Fr_{X \to} - Fr_{X \to Y} + Fr_{Y \to} - Fr_{Y \to X}. \tag{4.26}$$

It should be pointed out that the relationships defined in (4.25) and (4.26) are true irrespective of the independence assumption of components. This expression can be simplified later on in Section 4.4 using the concept of frequency balance.

An example of how to calculate frequency both directly and the frequency union of disjoint sets is given next.

Example 4.3 Consider a system consisting of three two-state generators, each having a capacity of 10 MW. The two-state generator model and the eight-state system model are shown in Figures 4.7 and 4.8, respectively. Let us say we want to find the probability of an event with the capacity less than or equal to 10 MW and the frequency of encountering and exiting that event.

The first step is to identify all states that form the subset indicating this event. The probability of this subset can be easily found by adding the probabilities of

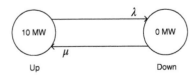

Figure 4.7 A two-state component model of a single generator.

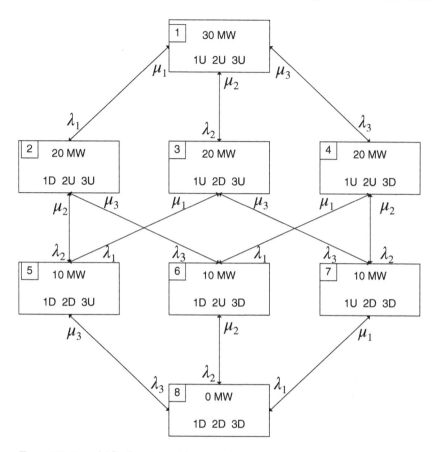

Figure 4.8 A model for three two-state generators.

these states. For determining the frequency, it is convenient to draw a boundary around these states, as shown in Figure 4.9.

Let $E = \{(\text{Cap} \le 10)\}$; we can then find the expected transition rate across this boundary as follows:

$$p_E = p_5 + p_6 + p_7 + p_8 \tag{4.27}$$

$$Fr_{\to E} = p_2(\lambda_2 + \lambda_3) + p_3(\lambda_1 + \lambda_3) + p_4(\lambda_1 + \lambda_2) \tag{4.28}$$

$$Fr_{E\to} = p_5(\mu_1 + \mu_2) + p_6(\mu_1 + \mu_3) + p_7(\mu_2 + \mu_3). \tag{4.29}$$

Alternatively, we can use frequency of union of disjoint sets to find the frequency of encoutering and exiting E. Let $E = X \cup Y$, where $X = \{(\text{Cap} = 10)\}$ and $Y = \{(\text{Cap} = 0)\}$. Since $X \cap Y = \emptyset$, we can use (4.25) and (4.26).

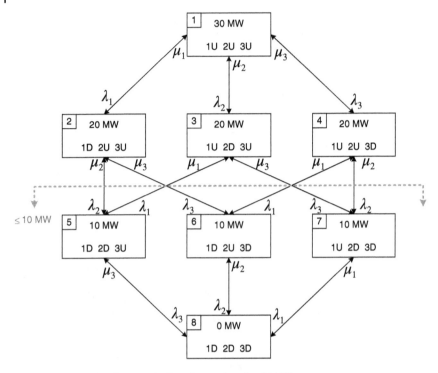

Figure 4.9 Boundary for capacity less than or equal to 10 MW.

Substitute the following:

$$Fr_{\to X} = p_2(\lambda_2 + \lambda_3) + p_3(\lambda_1 + \lambda_3) + p_4(\lambda_1 + \lambda_2) + p_8(\mu_1 + \mu_2 + \mu_3),$$
$$Fr_{Y \to X} = p_8(\mu_1 + \mu_2 + \mu_3),$$
$$Fr_{\to Y} = p_5\lambda_3 + p_6\lambda_2 + p_7\lambda_1,$$
$$Fr_{X \to Y} = p_5\lambda_3 + p_6\lambda_2 + p_7\lambda_1,$$

into (4.25), and we can find the frequency of entering $E = X \cup Y$, the same as in (4.28).

Similarly, we can substitute $Fr_{X \to Y}, Fr_{Y \to X}$ and:

$$Fr_{X \to} = p_5(\mu_1 + \mu_2 + \lambda_3) + p_6(\mu_1 + \mu_3 + \lambda_2) + p_7(\mu_2 + \mu_3 + \lambda_1),$$
$$Fr_{Y \to} = p_8(\mu_1 + \mu_2 + \mu_3),$$

into (4.26), and we can find the frequency of exiting $E = X \cup Y$, the same as in (4.29).

In this example, we find the *cumulative* probability of an event with capacity less than 10 MW. Likewise, we call the frequency of exiting an event with capacity equal to or less than 10 MW *cumulative frequency*.

It should be particularly noted that the cumulative frequency, unlike the cumulative probability, cannot be obtained by summing the frequencies of individual states, as only those frequencies that cross the boundary count towards the cumulative frequency.

4.4 Concept of Frequency Balance

The idea of frequency balance seen in the Example 4.2 is generalized in this section. We begin with the concept of frequency balance on any state i in the state space S and then extend the concept of frequency balance to any subset $X, Y \subset S$.

4.4.1 Frequency Balance of a State

In steady state or average behavior, the frequency of encountering a state equals the frequency of exiting the state. In other terms, in steady state there is a balance between two disjoint subsets of the state space, i.e., $\forall i \in S$,

$$Fr_{\to\{i\}} = Fr_{\{i\}\to}, \tag{4.30}$$

where $Fr_{\to\{i\}}$ and $Fr_{\{i\}\to}$ represent frequency of entering state i and frequency of exiting state i, respectively.

From this point onward, we will denote a frequency of a state as $Fr_{\{i\}}$ and remove the direction of the frequency. This simple idea can be used to write the state equations and find the steady state probability of any states in the system, as we can see from the following example.

Example 4.4   Consider the same system as in Example 4.1. We can use the frequency balance concept in (4.30) to calculate the state probabilities. We can write two equations, one describing the frequency balance of state 1, and other describing the frequency balance of state 2.

The steady state equation for state 1 can be written as (4.31):

$$Fr_{\to\{1\}} = Fr_{\{1\}\to}. \tag{4.31}$$

Now state 1 is communicating with state 2. Therefore, equation (4.31) can be written as

$$Fr_{\{2\}\to\{1\}} = Fr_{\{1\}\to\{2\}}. \tag{4.32}$$

That is,

$$p_2\mu = p_1\lambda. \tag{4.33}$$

Likewise, the frequency balance of state 2 can be found from (4.34):

$$Fr_{\to\{2\}} = Fr_{\{2\}\to}. \tag{4.34}$$

That is,

$$Fr_{\{1\}\to\{2\}} = Fr_{\{2\}\to\{1\}}, \tag{4.35}$$

$$p_1\lambda = p_2\mu. \tag{4.36}$$

Note that both (4.33) and (4.36) are exactly the same. We need one more linearly independent equation in order to solve for the steady state probabilities. In this example, we can use

$$p_1 + p_2 = 1. \tag{4.37}$$

Then, $p_1 = \frac{\mu}{\mu+\lambda}$ and $p_2 = \frac{\lambda}{\mu+\lambda}$.

From the previous example, we can see that when using the concept of frequency balance, n equations can be written for n system states. These equations are, however, not linearly independent, so only $n-1$ of these can be used. The nth equation required is the total probability equation.

$$\sum_{i=1}^{n} p_i = 1 \tag{4.38}$$

In general, the n equations obtained can be solved to find the steady state probabilities. It should be noted that if the components are independent, the system state probabilities can be simply obtained from the component state probabilities by using the multiplication rule of probabilities.

4.4.2 Frequency Balance of a Set

The concept of frequency balance is also true for any set in the state space. For any set $Y \subset S$,

$$Fr_{\to Y} = Fr_{Y\to}. \tag{4.39}$$

In this case, we also denote a frequency of a state as Fr_Y and remove the direction of the frequency from this point onward. Equivalently, we can write

$$Fr_Y = Fr_{\bar{Y}}. \tag{4.40}$$

Note that both the frequency balance of a state and a set are found between two disjoint subsets that occupy the whole state space, i.e., in this case, $Y \cup \bar{Y} = S$. The frequency between any pair of states or any pair of disjoint subsets are *not* necessarily balanced.

Figure 4.10 A frequency-balanced component.

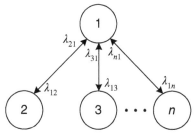

For a component whose the frequency between states and subsets are balanced, we call it *Balanced Frequency (BF) Components*. The components are *BF* if and only if (4.41) is true:

$$Fr_{X \to Y} = Fr_{Y \to X} \tag{4.41}$$

for every $X, Y \subset S$ and $X \cap Y = \emptyset$. This means that there is a frequency balance between every pair of system states or any pair of mutually exclusive subsets.

Examples of multistate *BF* components and non-*BF* components are shown in Figures 4.10 and 4.11, respectively.

As a result of (4.39), for any $X \subset S$ and $X \cap Y = \emptyset$, we have from (4.19) and (4.21):

$$Fr_{X \to Y} + Fr_{\bar{X} \backslash Y \to Y} = Fr_{Y \to X} + Fr_{Y \to \bar{X} \backslash Y}. \tag{4.42}$$

If the components are frequency balanced, we have

$$Fr_{\bar{X} \backslash Y \to Y} = Fr_{Y \to \bar{X} \backslash Y}. \tag{4.43}$$

Equation (4.43) will be used to find the frequency of union of disjoint events in a more efficient manner. Recall the frequency of union of events given in (4.25) and (4.26); we can write shortly as (4.44):

$$Fr_{X \cup Y} = Fr_X - Fr_{Y \to X} + Fr_Y - Fr_{X \to Y}. \tag{4.44}$$

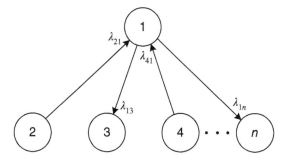

Figure 4.11 A non-frequency-balanced component.

From (4.19), $Fr_Y = Fr_{X \to Y} + Fr_{\bar{X} \backslash Y \to Y}$, we have

$$Fr_{X \cup Y} = Fr_X - Fr_{Y \to X} + Fr_{\bar{X} \backslash Y \to Y}. \tag{4.45}$$

When the components are frequency balanced, we can use (4.43); then a more efficient relationship can be derived [2] as follows:

$$Fr_{X \cup Y} = Fr_X - Fr_{Y \to X} + Fr_{Y \to \bar{X} \backslash Y}. \tag{4.46}$$

Equation (4.46) allows the flexibility to calculate the frequency in a more efficient manner when there are fewer states in Y compared to the number of states in $\bar{X} \backslash Y$. The calculation of frequency will have fewer terms when the direction of the frequency is from Y than that from $\bar{X} \backslash Y$. However, (4.46) is true only if the components are frequency balanced. We illustrate the use of (4.46) in the following example.

Example 4.5 Consider the same system as shown in Example 4.3. We can find a frequency of the events with generator two being down, $E = \{3, 5, 6, 7, 8\}$, using both (4.44) and (4.46).

Let $X = \{5, 6, 7, 8\}$, $Y = \{3\}$; then, using (4.44), Fr_E can be found from the following:

$$Fr_X = p_5(\mu_1 + \mu_2) + p_6(\mu_1 + \mu_3) + p_7(\mu_2 + \mu_3),$$
$$Fr_{Y \to X} = p_3(\lambda_1 + \lambda_3),$$
$$Fr_{\bar{X} \backslash Y \to Y} = p_1(\lambda_2) + p_2(0) + p_3(0).$$

If we use (4.46), we need to find $Fr_{Y \to \bar{X} \backslash Y}$ instead of $Fr_{\bar{X} \backslash Y \to Y}$ as follows:

$$Fr_{Y \to \bar{X} \backslash Y} = p_3(\mu_2).$$

Note that when we use (4.46), we only need to consider the transition from state $\{3\}$ to states $\{1, 2, 4\}$, whereas when we use (4.44), we need to find the transitions from three states, $\{1, 2, 3\}$, to state $\{3\}$.

It is common to see the discrete capacities of both generation and transmission lines in electric power systems similar to both Examples 4.3 and 4.5. This leads us to a more efficient algorithm to calculate frequency for this special case in the next section.

4.4.3 Special Results for Discrete Capacity Systems

As an interesting application, equations (4.44) and (4.46) are used to develop a more efficient algorithm for a limited application in power generation system modeling. Typically, the generating capacity is discrete, similar to Example 4.3, as shown in Figure 4.8. A capacity-outage probability and frequency table is often used in reliability evaluation of electric power generation systems. This table consists of the capacity associated with state i, C_i, cumulative probability and frequency, i.e., $Pr(\{\text{Cap} \leq C_i\})$ and $Fr(\{\text{Cap} \leq C_i\})$, associated with each level of capacity.

Let us assume that there are n capacity states with C_i indicating the capacity associated with state i. For example, there are four capacity states, namely, 0, 10, 20 and 30 MW for the system shown in Figure 4.8. Let us further assume that the states are arranged in the descending order of capacity, i.e., $C_i < C_{i-1}$. The objective here is to compute cumulative probability and frequency associated with each level of capacity recursively.

From $\{\text{Cap} \leq C_i\} = \{\text{Cap} \leq C_{i-1}\} \cup \{\text{Cap} = C_i\}$, let

$$f_{i-} = Fr(\{\text{Cap} = C_i\} \rightarrow \{\text{Cap} \leq C_{i-1}\}), \tag{4.47}$$

$$f_{+i} = Fr(\{\text{Cap} > C_i\} \rightarrow \{\text{Cap} = C_i\}), \tag{4.48}$$

$$f_{i+} = Fr(\{\text{Cap} = C_i\} \rightarrow \{\text{Cap} > C_i\}), \tag{4.49}$$

where

f_{i-} Frequency of transiting from C_i to states with lower capacity.
f_{+i} Frequency of encountering C_i from states with higher capacity.
f_{i+} Frequency of exiting C_i to states with higher capacity.

Equation (4.44) can be used to produce a recursive formula as follows:

$$Fr(\{\text{Cap} \leq C_i\}) = f_{i-1} - f_{i-} + f_{+i}. \tag{4.50}$$

Equation (4.50) shows that we can find a frequency of a capacity state in any states recursively. If the system is *BF*, we can use (4.46) as follows:

$$Fr(\{\text{Cap} \leq C_i\}) = f_{i-1} - f_{i-} + f_{i+}. \tag{4.51}$$

Equation (4.50) is general and does not need statistical independence of components or satisfaction of (4.41), whereas (4.51) requires that the components be independent and frequency balanced. Equation (4.51) is, however, much easier to implement because f_{i+} is simpler to calculate than f_{+i}, especially, for the case where the (4.51) can be shown equivalent to the recursive relationship presented in [3]. Other applications of some limited versions of (4.44) can be found in [4] [5] [6].

The use of (4.46) has been restricted to generation systems consisting of two-state components when the components are independent and frequency balanced. When the components are not frequency balanced, even if all of them are independent in the state space, we can not find the frequency using a simple recursive formula. In the next section, we circumvent this problem using the concept of forced frequency balance. The application includes a generating unit shown in Figure 4.11, which has three component states and is not frequency balanced.

4.4.4 Forced Frequency Balance

It is shown in the previous section that (4.46) is superior to (4.44) because it is comparatively easier to calculate $Fr_{Y \to \bar{X} \backslash Y}$, as generally Y has only a few terms to be considered, than $\bar{X} \backslash Y$. However, the problem with (4.46) is that it requires all components to be independent and frequency balanced. This limitation can be overcome using the concept of forced frequency balance [7], which is applied between subsets of the state space and is explained below.

This concept is derived for the discrete capacity systems, which are the same as the system considered in Section 4.4.3. Let us first begin with a set of definitions.

Definition 4.1 *A* discrete capacity component *is the component that a definite capacity can be assigned to each state* k *of the component state.*

Definition 4.2 *A system is* discrete capacity system *when a definite capacity can be assigned to each state* i *of the system, and the capacity of a system state is the sum of capacities of component states comprising that system state, i.e.,*

$$C_i = \sum_{\forall c} C_{ic_k}, \tag{4.52}$$

where

C_i *Capacity assigned to system state* i.
C_{ic_k} *Capacity of component* c *in component state* k *in system state* i.

Definition 4.3 *An* orientation of component states *is the direction from component state* c_k *to* c_l *in the system state* i. *It is positive if* $C_{ic_k} > C_{ic_l}$ *and is negative if* $C_{ic_k} < C_{ic_l}$.

Definition 4.4 *An* orientation of subsets of system states *is the direction from* $\{C_i\}$ *to* $\{C_j\}$, $\{C_i\}, \{C_j\} \subset S$. *It is positive if* $min_j\{C_j\} > max_i\{C_i\}$, *and is negative if* $max_j\{C_j\} < min_i\{C_i\}$.

A discrete-capacity component in system state i is *forced frequency-balanced* in the positive direction if, for the purpose of frequency computation, $\lambda_{c_k c_l}$ is replaced by (4.53):

$$\alpha_{c_k c_l} = \frac{Fr_{\{c_k\} \to \{c_l\}}}{p_{c_k}}, \forall k, l \text{ such that } C_{ic_k} > C_{ic_l}, \tag{4.53}$$

where

$\lambda_{c_k c_l}$ Actual transition rate of a component c from component state k to l.
$\alpha_{c_k c_l}$ Fictitious transition rate of a component c from component state k to l.

$Fr_{c_k \rightarrow c_l}$ Frequency of a component moving from state k to state l.
p_{c_k} Steady state probability of a component c being in component state k.

The process of forcing frequency balance onto component states in the positive direction consists of the following steps:

1. Compute component-state probabilities with the real transition rates $\lambda_{c_k c_l}$.
2. From each state c_i, compute fictitious transition rate $\alpha_{c_k c_l}$ for $C_{ic_k} > C_{ic_l}$.
3. Do not alter transition rates from c_i in the positive direction.

Similarly, a component is forced frequency balanced in the negative direction by replacing $\lambda_{c_k c_l}$ with (4.54):

$$\alpha_{c_k c_l} = \frac{Fr_{\{c_k\} \rightarrow \{c_l\}}}{p_{c_k}}, \forall k, l \text{ such that } C_{ic_k} < C_{ic_l}. \tag{4.54}$$

We now state the following theorem. This theorem is used to calculate the frequency of union of disjoint events in the state space when all components are independent but some are non frequency-balanced.

Theorem 4.1 *For discrete capacity systems consisting of independent discrete capacity components, for any* $X, Y \subset S, X \cap Y = \emptyset$,

$$Fr_{(X \cup Y)} = \begin{cases} Fr_X - Fr^-_{Y \rightarrow X} + Fr_{Y \rightarrow \bar{X} \backslash Y} & ,X > Y \\ Fr_X - Fr^+_{Y \rightarrow X} + Fr_{Y \rightarrow \bar{X} \backslash Y} & ,X < Y, \end{cases} \tag{4.55}$$

where the superscript of Fr *indicates the direction of forced frequency balance of the non frequency-balanced components in* Y

The proof of this theorem can be found in [7], and some interesting applications of this concept can be found in [8] [9]. The following example illustrates some of these ideas.

Example 4.6 Consider a system of two three-state generating units. The state-transition diagram of a single unit is shown in Figure 4.12, which is a

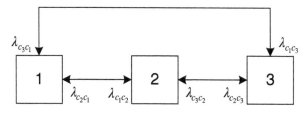

Figure 4.12 Transition diagram of a three-state generating unit.

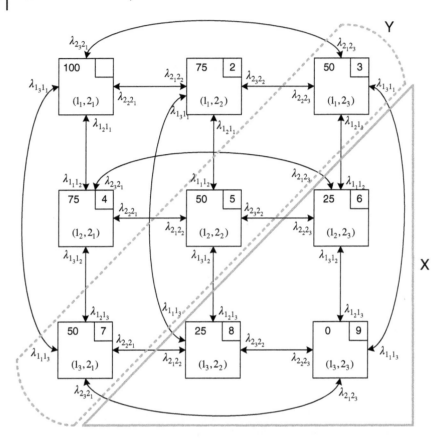

Figure 4.13 A system of two three-state generating units.

frequently used model for a large generating unit [7]. The system state-transition diagram is shown in Figure 4.13, where the number in the top-right corner is system state i, in the top-left corner is C_i and the numbers in parentheses are the component states c_k.

In this example, we have two components, $c \in \{1, 2\}$, and the component states are up state, derated state and down state, represented by the numerical value 1, 2 and 3, respectively. The component state capacities are given as follows. For any state i, if the component is in up state, $C_{i1_1} = C_{i2_1} = 50$ MW. If the component is in derated state, $C_{i1_2} = C_{i2_2} = 25$ MW. If the component is in derated state, $C_{i1_3} = C_{i2_3} = 0$ MW.

The state space of this system is $\{1, 2, \dots, 9\}$, where the capacities of each system state i is 0, 25, 50, 75 and 100 MW. Let $X = \{6, 8, 9\}$, i.e., capacity

$\leq 25\,\mathrm{MW}$ and $Y = \{3, 5, 7\}$, i.e., capacity of 50 MW. We can find $Fr_{X \cup Y}$ by using Theorem 4.1.

Since the direction from $X \to Y$ is positive,

$$Fr_X = \sum_{i \in X} p_i \Big(\sum_{j \in \bar{X}} \lambda_{ij} \Big)$$

$$= p_6(\lambda_{1_2 1_1} + \lambda_{2_3 2_2} + \lambda_{2_3 2_1}) + p_8(\lambda_{1_3 1_1} + \lambda_{1_3 1_2} + \lambda_{2_2 2_1})$$
$$+ p_9(\lambda_{1_3 1_1} + \lambda_{2_3 2_1})$$

$$Fr^-_{Y \to X} = p_3(\alpha_{1_1 1_2} + \alpha_{1_1 1_3}) + p_5(\alpha_{1_2 1_3} + \alpha_{2_2 2_3}) + p_7(\alpha_{2_1 2_3} + \alpha_{2_1 2_2})$$

$$Fr_{Y \to \bar{X} \backslash Y} = p_3(\lambda_{2_3 2_1} + \lambda_{2_3 2_2}) + p_5(\lambda_{1_2 1_1} + \lambda_{2_2 2_1}) + p_7(\lambda_{1_3 1_1} + \lambda_{1_3 1_2}).$$

Since all the component transitions in $Fr^-_{Y \to X}$ have positive direction, we can substitute $\alpha_{c_k c_l}$ using (4.53):

$$\alpha_{1_1 1_2} = \frac{p_{1_2} \lambda_{1_2 1_1}}{p_{1_1}}$$

$$\alpha_{1_1 1_3} = \frac{p_{1_3} \lambda_{1_3 1_1}}{p_{1_1}}$$

$$\alpha_{1_2 1_3} = \frac{p_{1_3} \lambda_{1_3 1_2}}{p_{1_2}}$$

$$\alpha_{2_2 2_3} = \frac{p_{2_3} \lambda_{2_3 2_2}}{p_{2_2}}$$

$$\alpha_{2_1 2_3} = \frac{p_{2_3} \lambda_{2_3 2_1}}{p_{2_1}}$$

$$\alpha_{2_1 2_2} = \frac{p_{2_2} \lambda_{2_2 2_1}}{p_{2_1}}.$$

Since the components are independent, we can write (4.56):

$$p_i = \prod_{\forall c} p_{c_k}. \tag{4.56}$$

Using (4.56), we can write

$$p_3 = p_{1_1} p_{2_3}$$
$$p_5 = p_{1_2} p_{2_2}$$
$$p_7 = p_{1_3} p_{2_1}.$$

We can find

$$Fr^-_{Y \to X} = p_{1_1} p_{2_3} \left(\frac{p_{1_2} \lambda_{1_2 1_1}}{p_{1_1}} + \frac{p_{1_3} \lambda_{1_3 1_1}}{p_{1_1}} \right)$$

$$+ p_{1_2} p_{2_2} \left(\frac{p_{1_3} \lambda_{1_3 1_2}}{p_{1_2}} + \frac{p_{2_3} \lambda_{2_3 2_2}}{p_{2_2}} \right)$$

$$+ p_{1_3} p_{2_1} \left(\frac{p_{2_3} \lambda_{2_3 2_1}}{p_{2_1}} + \frac{p_{2_2} \lambda_{2_2 2_1}}{p_{2_1}} \right).$$

Then, we can substitute the following equation to find the frequency of $X \cup Y$:

$$Fr^-_{Y \to X} = p_6 (\lambda_{1_2 1_1} + \lambda_{2_3 2_2}) + p_8 (\lambda_{1_3 1_2} + \lambda_{2_2 2_1}) + p_9 (\lambda_{1_3 1_1} + \lambda_{2_3 2_1}).$$

Finally,

$$Fr_{X \cup Y} = p_3 (\lambda_{2_3 2_1} + \lambda_{2_3 2_2}) + p_5 (\lambda_{1_2 1_1} + \lambda_{2_2 2_1}) + p_6 \lambda_{2_3 2_1}$$

$$+ p_7 (\lambda_{1_3 1_1} + \lambda_{1_3 1_2}) + p_8 \lambda_{1_3 1_1}.$$

We can easily verify this result with the state-transition diagram in Figure 4.13.

It should be noted that (4.46) is a special case of (4.55), which has been explicitly or implicitly used in [6] [10] [11]. For systems consisting of multistate components, some corrective terms have been used in [11] along with (4.46). The forced frequency balance shown in this section allows us to calculate the frequency of units of disjoint sets in a straightforward manner without using any corrective terms. If, however, components are forced frequency balanced in the negative direction, (4.54) should be used instead of (4.53).

As can be seen from Examples 4.3 and 4.6, the system state grows exponentially with number of components, but the number of capacity states is still manageable. One way to reduce number of states in the state space is to group states with the same characteristics; for example, in Example 4.3 we may reduce the state space to the states with capacities of 0, 10, 20 and 30 MW. This substantially reduces number of states from eight to four. We describe this idea in the next section.

4.5 Equivalent Transition Rate

Sometimes it is convenient to reduce the state space by combining states. The reduced state space can then be more conveniently combined with the models of the other subsystems.

Using the concept of frequency, in the steady state the *equivalent transition rate* from disjoint subsets X to Y, $X, Y \subset S$, and $X \cap Y = \emptyset$, can be obtained as

$$\lambda_{XY} = \frac{Fr_{X \to Y}}{p_X}. \tag{4.57}$$

The conditions governing the mergeability are discussed fully in [12], but an appreciation of these can be obtained by considering an example of two components.

Example 4.7 The state-transition diagram of two two-state components is shown in Figure 4.14, where λ_i and μ_i are the failure and repair rates of component *i*.

Let us say that we want to represent this four-state model by three states, a, b and c, as shown in the Figure 4.14. We can apply (4.57) to find equivalent transition rate as follows:

$$\lambda_{ab} = \frac{Fr_{a \to b}}{p_a} \tag{4.58}$$

$$= \frac{p_a(\lambda_1 + \lambda_2)}{p_a} \tag{4.59}$$

$$= \lambda_1 + \lambda_2 \tag{4.60}$$

$$\lambda_{ba} = \frac{Fr_{b \to a}}{p_b} \tag{4.61}$$

$$= \frac{p_2\mu_1 + p_3\mu_2}{p_2 + p_3}. \tag{4.62}$$

Similarly,

$$\lambda_{bc} = \frac{p_2\lambda_2 + p_3\lambda_1}{p_2 + p_3} \tag{4.63}$$

$$\lambda_{cb} = \mu_1 + \mu_2. \tag{4.64}$$

Figure 4.14 Three-state equivalent of four-state model.

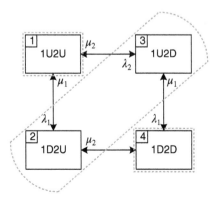

It can be seen that equivalent transition rates from a to b and b to c are independent of state probabilities and therefore can be readily applied in combining this reduced model with models of other subsystems. However, transition rates from b to a and b to c are functions of the state probabilities. These transition rates can be applied when dealing with steady state and when the subsystems being combined are independent. It can be seen that for the special case of identical components, where $\lambda_1 = \lambda_2 = \lambda$ and $\mu_1 = \mu_2 = \mu$, we have

$$\lambda_{ab} = 2\lambda$$
$$\lambda_{ba} = \mu$$
$$\lambda_{bc} = \lambda$$
$$\lambda_{cb} = 2\mu.$$

and these transition rates are independent of state probabilities and can be applied to both steady state and time-specific calculations.

Example 4.7 shows how to combine states in a state space in order to represent the state space in a more compact manner. The application of this approach is to efficiently determine the probability or frequency of encoutering an event with the same impact to the system.

As mentioned in Chapter 1, we are interested in evaluating measures related to failure states of the system—for example, failure probability and failure frequency. By grouping states in the state space, whose impact of each state represents the same charateristics of the system, the failure states can be easily detected and reliability can be effectively evaluated.

As the size of the state space grows, we may need an approach to calcaulate frequency index by reducing the size of the state space. This is achieved by conditional probability rule, which is used mainly to find probability of an event, as outlined in Chapter 2.

Unlike the way we calculate probability, the process of determining frequency involves these transitions, and we need to make sure that the impact of the transition will be as we expected. For example, if the system is currently in a failure state, when one extra component fails it should not bring the system to a good working state.

This concept is called *coherence* and is explained in the following section. We then later present how we determine the frequency of an event, called *conditional frequency*, in Section 4.7 with the assumption of coherency.

4.6 Coherence

Coherence is a property that relates to the status of the system when changing states. We usually use this property and apply it to the change between failure and success states. It can be described simply as follows:

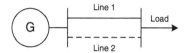

Figure 4.15 Reliability coherence when transmission line 2 fails.

- If the system is in a failure state with some components healthy and some components failed, then for a reliability coherent system failure of a healthy component will not lead to system success.
- If the system is in success state with some components healthy and some components failed, then for a reliability coherent system repair/restoration of a failed component will not lead to system failure.

In simple terms this concept can be described as follows. In a reliability coherent system, component degradation will not lead to system improvement and component improvement will not lead to system degradation. This can be further clarified by the following examples.

Example 4.8 Consider the system shown in Figure 4.15.

Transmission line 2 in the system is currently in the failure state. In this example, assuming that the system is in the success state, the system is said to be reliability coherent if the repair of line 2 does not result in system failure.

Example 4.9 Consider a system shown in Figure 4.16.

Generator 2 in the system is currently in the failure state. In this example, assuming that the system is in the failed state, the system is said to be reliability coherent if the failure of a working component, such as a generator 1 or line 1, does not result in system success.

A comparison between state space of coherent and non coherent systems is illustrated by Figure 4.17. Both state spaces are divided into two disjoint subsets conditioned on the change in status of component k.

Let us denote Y as a set of failure states and $S \setminus Y$ as a set of success states and component k is failed. If the system is coherent, the states in Y cannot change their status from success to failure when component k is repaired. Similarly, the

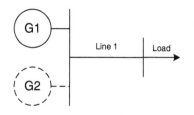

Figure 4.16 Reliability coherence when generator 2 fails.

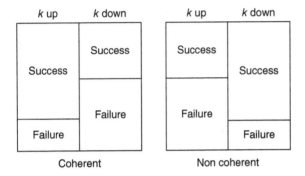

Figure 4.17 A state space representation when a two-state component k changes its status in (a) a coherent system and (b) a non coherent system.

states in $S \setminus Y$ cannot change their status from failure to success when component k fails. Let K be an event that component k is in up state and $\bar{K}$ be an event that component k is in down state; we can compare the conditional probabilities of the two cases as follows if the system is coherent:

$$P\{Y|\bar{K}\} \geq P\{Y|K\} \qquad (4.65)$$

and

$$P\{(S \setminus Y)|K\} \geq P\{(S \setminus Y)|\bar{K}\}. \qquad (4.66)$$

Equation (4.65) simply means that the failure probability of the system when component k fails is higher than that when component k is working. Likewise, (4.66) tells us that the success probability when component k is working is higher when the component k is in good condition than when it deteriorates.

It will be evident in the next section why coherence is an important condition of the system in order to make a conditional frequency calculation possible.

4.7 Conditional Frequency

We present an efficent approach to calculate frequency index using conditional probability rule, which is used mainly to find probability of an event, as outlined in Chapter 2. This concept is a very powerful tool for evaluting probability of an event, as we can divide the state space into disjoint sets and evaluate probability of events over the disjoint sets separately. This concept allows the calculation to be made over a smaller and managable size of the state space.

In this section, we are interested in finding frequency of encoutering a set Y in the state space S, $Y \subset S$. Recall (4.17) and (4.18) give the frequency of entering and exiting a set Y, which are shown again below:

$$Fr_{\to Y} = \sum_{i \in S \setminus Y} (p_i \sum_{j \in Y} \lambda_{ij})$$

$$Fr_{Y \to} = \sum_{i \in Y} (p_i \sum_{j \in S \setminus Y} \lambda_{ij}).$$

Since the focus of reliability analysis is to evaluate frequency of entering failure states, set Y generally represents the failure set. Usually the number of failure states is smaller than that of success states. This means that it is easier to calculate the frequency of exiting the failure set than the frequency of encountering the failure set because only failure state probabilities are needed. Since we know from Section 4.4.1 that frequency of a set is always balanced at the steady state, we can write only Fr_Y and use (4.18) instead of (4.17).

The following assumptions are made throughout this section:

1. The system is repairable and composed of independent components.
2. Each component is represented by a two-state Markov model.
3. The system is coherent.

For a n-component independent system, the probability of a system state i is easily found from multiplication of component state probabilities, as given in (4.67):

$$p_i = \prod_{k=1}^{n} P\{c_i(k)\}, \tag{4.67}$$

where $P(\{c_i(k)\})$ is the probability of the status of component k, $c_i(k)$ status in system state i and $k \in \{1, 2, \ldots, n\}$.

Another important property that follows from the independence assumption is that a transition between any pair of states i, j can only be made by changing a component state of one component at a time. Since each component can only assume two states, up or down, we let K be an event that a component k is working and $\bar{K}$ be an event that the component k fails. We can write $Y = (Y|K) \cup (Y|\bar{K})$, and using (4.16),

$$Fr_{Y \to} = Fr_{[(Y|K) \cup (Y|\bar{K})] \to S \setminus Y}$$
$$= Fr_{Y|K \to S \setminus Y} + Fr_{Y|\bar{K} \to S \setminus Y}.$$

Similarly, we can write $S \setminus Y = [(S|K) \setminus Y] \cup [(S|\bar{K}) \setminus Y]$, and again using (4.16),

$$Fr_{Y \to} = Fr_{Y|K \to (S|K) \setminus Y} + Fr_{Y|\bar{K} \to (S|K) \setminus Y} + Fr_{Y|K \to (S|\bar{K}) \setminus Y} + Fr_{Y|\bar{K} \to (S|\bar{K}) \setminus Y} \tag{4.68}$$

Equation (4.68) gives us the conditional frequency of the set Y conditioned on the status of component k. The first and the last term of (4.68) are the frequency of an event Y given that the component k is working and the frequency of an event Y given that the component k fails respectively. We now investigate carefully the second and the third terms.

Consider both $Fr_{Y|\bar{K}\to(S|K)\backslash Y}$ and $Fr_{Y|K\to(S|\bar{K})\backslash Y}$; let us substitute $(S|K) \backslash Y = (S \backslash Y)|K$ and $(S|\bar{K}) \backslash Y = (S \backslash Y)|\bar{K}$; then, we can write the second and the third term in (4.68) as follows:

$$Fr_{Y|\bar{K}\to(S|K)\backslash Y} = Fr_{Y|\bar{K}\to(S\backslash Y)|K} \qquad (4.69)$$

$$Fr_{Y|K\to(S|\bar{K})\backslash Y} = Fr_{Y|K\to(S\backslash Y)|\bar{K}}. \qquad (4.70)$$

Equations (4.69) and (4.70) are the frequencies of moving from states in Y to $S \backslash Y$ when the component k changes its status from K to $\bar{K}$, or vice versa.

When the system is coherent, it means that when the component is repaired, i.e., changes its state from $\bar{K}$ to K, it does not make the system change from failure (Y) to success $(S \backslash Y)$. Since this transition can never exist for a coherent system, it simply implies that (4.70) is zero. Equation (4.68) can then be simplified as follows:

$$Fr_{Y\to} = Fr_{Y|K\to(S\backslash Y)|K} + Fr_{Y|\bar{K}\to(S\backslash Y)|\bar{K}} + Fr_{Y|\bar{K}\to(S\backslash Y)|K}. \qquad (4.71)$$

We have in (4.71) the conditional frequency of set Y conditioned upon the status of component k. As we can see from this expression, it contains three terms, the first being the frequency given that the component k is up, the second being the frequency given that the component k is down and the last frequency as a result of the change in component k's state.

We can also derive an alternative expression for conditional frequency given in (4.71). Using (4.17) instead of (4.18), we can write (4.72):

$$Fr_{\to Y} = Fr_{(S\backslash Y)|K\to Y|K} + Fr_{(S\backslash Y)|\bar{K}\to Y|\bar{K}} + Fr_{(S\backslash Y)|K\to Y|\bar{K}}. \qquad (4.72)$$

Equation (4.72) assumes coherency since $Fr_{Y|K\to(S\backslash Y)|\bar{K}} = 0$, which implies that when the component k changes its state from working condition to failure, it cannot make the system state move from failure to success state.

We can see that for both (4.71) and (4.72) the first two terms are the frequency given that the component k is either up or down. *This allows the frequency calculation to be simpler since the state space is now divided into two smaller disjoint sets.* We now discuss how to calculate the last term.

Since the frequency is the expected transition and is found from multiplication between the probability and the transition rate, we need to find the probability of states that transit from Y to $S \backslash Y$ when the component k is repaired.

Let us first describe the failure set in Figure 4.18 when the component k fails, $Y|\bar{K}$, consisting of two parts: (1) the states that are failure states when component k is in up state and (2) the states which are success states, given that the

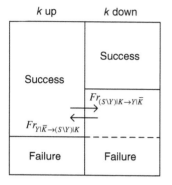

k up	k down

Success

Success

$Fr_{(S\backslash Y)|K\to Y|\bar{K}}$ →

← $Fr_{Y|\bar{K}\to(S\backslash Y)|K}$

Failure Failure

Figure 4.18 Failure set description in the state space when one two-state component changes its status.

component k is in up state. It is of our interest to find the probability of states in part (2).

Note that coherency implies that *all* the failure states in Y remain in the failure state when the component k fails. This means that the conditional probability of these states, whether the component k fails or is repaired, remains the same, i.e., the probability of part (1) is $P\{Y|K\}$. This implies that the probability of part (2) can be found by subtracting from the failure probability when the component k fails the failure probability when the component k repaired, which is given in (4.73):

$$P\{Y|\bar{K}\} - P\{Y|K\}. \tag{4.73}$$

Using the probability in (4.73), we can find the frequency of transitioning from failure states in $Y|\bar{K}$ to success states in $Y|K$ by multipling (4.73) with the probability of component k being in the failure state and its transition rate from down to up state.

Let us call the frequency of failure as a result of change in the state of component k from down to up state $Fr_{Y_k\to}$. Denote the probability of k being in the failure state as $p_{\bar{k}}$ and the transition rate from down to up state of component k as μ_k; we then have (4.74):

$$Fr_{Y_k\to} = Fr_{Y|\bar{K}\to(S\backslash Y)|K} = (P\{Y|\bar{K}\} - P\{Y|K\}) \times p_{\bar{k}}\mu_k. \tag{4.74}$$

Equivalently, we can also find the frequency of failure as a result of the change in state of component k from up to down state as $Fr_{\to Y_k}$.

$$Fr_{\to Y_k} = Fr_{(S\backslash Y)|K\to Y|\bar{K}} = (P\{(S\backslash Y)|K\} - P\{(S\backslash Y)|\bar{K}\}) \times p_k\lambda_k, \tag{4.75}$$

where p_K is the probability of the component k in up state, $p_K = 1 - p_{,\bar{K}}$ and λ_k is the transition rate of the component k from up state to down state.

We can write conditional frequency of set Y using (4.71) as follows:

$$Fr_{Y\to} = Fr_{Y|K\to(S\backslash Y)|K} + Fr_{Y|\bar{K}\to(S\backslash Y)|\bar{K}} + Fr_{Y_k\to}. \tag{4.76}$$

Similarly, if we use (4.72) we can write conditional frequency of set Y as follows:

$$Fr_{\to Y} = Fr_{(S \setminus Y)|K \to Y|K} + Fr_{(S \setminus Y)|\bar{K} \to Y|\bar{K}} + Fr_{\to Y_k}. \tag{4.77}$$

Now, if we let $E_1 = S|K$ and $E_2 = S|\bar{K}$, we can apply the same concept to these two sets on all other components $m, m \in \{1, 2, \ldots, n\} \setminus \{k\}$. This allows us to represent the frequency of set Y compactly as (4.78):

$$Fr_{Y \to} = \sum_{k=1}^{n} Fr_{Y_k \to} \tag{4.78}$$

$$= \sum_{k=1}^{n} (P\{Y|\bar{K}\} - P\{Y|K\}) \times p_{\bar{k}} \mu_k \tag{4.79}$$

when we use (4.76).

Alternatively, we can use (4.77); we then have

$$Fr_{\to Y} = \sum_{k=1}^{n} Fr_{\to Y_k} \tag{4.80}$$

$$= \sum_{k=1}^{n} (P\{(S \setminus Y)|K\} - P\{(S \setminus Y)|\bar{K}\}) \times p_k \lambda_k. \tag{4.81}$$

Equations (4.78) and (4.80) give us conditional frequency of set Y from summation of frequency of an individual component k changing its status, one at a time.

We now illustrate the concept of conditional frequency using Example 4.10.

Example 4.10 In this example, Figure 4.19 represents a state space conditioned on a component k.

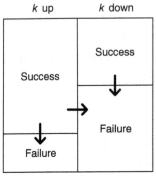

Coherent

Figure 4.19 A state space representation of state i when component k changes status.

It is clear in this example that frequency of failure of set Y comprises three main components.

In this section we observe that both (4.71) and (4.72) are the same due to the frequency-balanced concept. However, this is only true for steady state. In the following section the concept of time-specific state probabilities seen from Chapter 3 is extended to calculate *time-specific frequency*.

4.8   Time-Specific Frequency

We can calculate the time-specific frequency of exiting subset Y by substituting time-specific probability instead of steady state probability in (4.17):

$$Fr_{Y \to}(t) = \sum_{i \in Y} p_i(t)(\sum_{j \in S \backslash Y} \lambda_{ij}), \tag{4.82}$$

where $p_i(t)$ is the probability of system state i at time t.

Similarly, we can find the time specific frequency of entering subset Y as follows:

$$Fr_{\to Y}(t) = \sum_{i \in S \backslash Y} p_i(t)(\sum_{j \in Y} \lambda_{ij}). \tag{4.83}$$

In time domain, frequencies of exiting Y and entering Y are not balanced, although in steady state they are. Therefore, (4.82) and (4.83) are not equal in the time domain. Likewise, (4.79) and (4.81) are also not equal, as shown below:

$$Fr_{Y \to}(t) = \sum_{k=1}^{n}(P\{Y|\bar{K};t\} - P\{Y|K;t\}) \times p_{\bar{k}}(t)\mu_k \tag{4.84}$$

and

$$Fr_{\to Y}(t) = \sum_{k=1}^{n}(P\{(S \backslash Y)|K;t\} - P\{(S \backslash Y)|\bar{K};t\}) \times p_k(t)\lambda_k \tag{4.85}$$

Let us investigate the difference between these two above equations. Note that $P\{Y|K\} + P\{(S \backslash Y)|K\} = 1$ and $P\{Y|\bar{K}\} + P\{(S \backslash Y)|\bar{K}\} = 1$; we can write:

$$P\{Y|K\} + P\{(S \backslash Y)|K\} = P\{Y|\bar{K}\} + P\{(S \backslash Y)|\bar{K}\}.$$

Equivalently, we have

$$P\{Y|K\} - P\{Y|\bar{K}\} = P\{(S \backslash Y)|\bar{K}\} - P\{(S \backslash Y)|K\}, \tag{4.86}$$

which is always true for both the time-specific domain and steady state. This means that the difference between (4.84) and (4.85) is the last term, which are $p_{\bar{k}}(t)\mu_k$ and $p_k(t)\lambda_k$, respectively.

Note that we can also find out the differences between the two frequencies. Recall from Chapter 3 the relationship between state probabilities and transition rate in the time-specific domain in (3.44) and (3.41). They are applied in this case of two-state component below:

$$\frac{d}{dt}\left[p_k(t) \quad p_{\bar{k}}(t)\right] = \left[p_k(t) \quad p_{\bar{k}}(t)\right] \times \begin{bmatrix} -\lambda_k & \lambda_k \\ \mu_k & -\mu_k \end{bmatrix}. \tag{4.87}$$

Let us consider the elements in the matrix shown in (4.87); we can write the following equations:

$$\frac{d}{dt}p_{\bar{k}}(t) = p_k(t)\lambda_k - p_{\bar{k}}(t)\mu_k \tag{4.88}$$

$$\frac{d}{dt}p_k(t) = -p_k(t)\lambda_k + p_{\bar{k}}(t)\mu_k. \tag{4.89}$$

Equations (4.88) and (4.89) give the important relationship between $p_{\bar{k}}(t)\mu_k$ and $p_k(t)\lambda_k$. This means that the frequency of these two terms are not balanced by an amount of $-\frac{d}{dt}p_{\bar{k}}(t)$ and $-\frac{d}{dt}p_k(t)$, respectively.

Another concept of time-specific frequency is the *interval frequency*. The interval frequency of success can be obtained from (4.90):

$$Fr(t_1, t_2)_{Y\rightarrow} = \int_{t_1}^{t_2} Fr(t)_{Y\rightarrow}. \tag{4.90}$$

Equivalently, the interval frequency of failure can be obtained from (4.91):

$$Fr(t_1, t_2)_{\rightarrow Y} = \int_{t_1}^{t_2} Fr(t)_{\rightarrow Y}. \tag{4.91}$$

We can see that the time-specific frequency is the same concept as the steady state frequency except that the probability used in the calculation is time-specific probability. We are now interested in finding how we can easily convert the probability of failure to frequency of success or frequency of failure and how we can convert the probability of success to frequency of success or failure, as outlined in the next section.

4.9 Probability to Frequency Conversion Rules

We start the analysis with the time-specific domain as a general case. Let $p_s(t)$ be the probability of system success and $p_f(t)$ be the probability of system failure. Then,

$$p_s(t) = \sum_{i \in S\backslash Y} p_i(t) \tag{4.92}$$

$$p_f(t) = \sum_{i \in Y} p_i(t), \tag{4.93}$$

where $p_i(t)$ is the probability of system state i and is found from (4.67).

From the independence assumption, we can write the following expression for conditional probability:

$$P\{Y|\bar{K};t\} = \frac{P\{Y \cap \bar{K};t\}}{P\{\bar{K};t\}} = \frac{\sum_{i\in Y_{\bar{k}}} p_i(t)}{p_{\bar{k}}(t)} \tag{4.94}$$

$$P\{Y|K;t\} = \frac{P\{Y \cap K;t\}}{P\{K;t\}} = \frac{\sum_{i\in Y_{k}} p_i(t)}{p_{k}(t)} \tag{4.95}$$

$$P\{(S \setminus Y)|K;t\} = \frac{P\{(S \setminus Y) \cap K;t\}}{P\{K;t\}} = \frac{\sum_{i\in(S\setminus Y)_{k}} p_i(t)}{p_{k}(t)} \tag{4.96}$$

$$P\{(S \setminus Y)|\bar{K};t\} = \frac{P\{(S \setminus Y) \cap \bar{K};t\}}{P\{\bar{K};t\}} = \frac{\sum_{i\in(S\setminus Y)_{\bar{k}}} p_i(t)}{p_{\bar{k}}(t)}, \tag{4.97}$$

where $Y_{\bar{k}}$ is a subset of failure event Y that a component k fails and Y_k is a subset of failure event Y that a component k is working, $Y_{\bar{k}}, Y_k \subset Y$. Likewise, $(S \setminus Y)_k$ is a subset of success event $S \setminus Y$ that a component k is working and $(S \setminus Y)_{\bar{k}}$ is a subset of success event $S \setminus Y$ that a component k fails, $(S \setminus Y)_k, (S \setminus Y)_{\bar{k}} \subset (S \setminus Y)$. $p_k(t)$ and $p_{\bar{k}}(t)$ are the probabilities that a component k is up and down at time t, respectively.

We can rewrite (4.84) as follows:

$$Fr(t)_{Y\rightarrow} = \sum_{k=1}^{n} \left(\frac{\sum_{i\in Y_{\bar{k}}} p_i(t)}{p_{\bar{k}}(t)} - \frac{\sum_{i\in Y_{k}} p_i(t)}{p_{k}(t)} \right) \times p_{\bar{k}}(t)\mu_k$$

$$= \sum_{k=1}^{n} \sum_{i\in Y_{\bar{k}}} p_i(t)\mu_k - \sum_{k=1}^{n} \sum_{i\in Y_{k}} \left(p_i(t) \frac{p_{\bar{k}}(t)\mu_k}{p_{k}(t)} \right).$$

This means that we can calculate the frequency of encountering success states from the weighted summation of failure probability of state i, $p_i(t)$ with some terms. Note that the first term denotes the summation of repair rates of all components that are down in state i, and the second term denotes the summation of all components that are up in state i.

Since $Y = Y_k \cup Y_{\bar{k}}$, we can write (4.84) compactly as follows:

$$Fr(t)_{Y\rightarrow} = \sum_{i\in Y} p_i(t) \left(\sum_{k\in \bar{K}_i} \mu_k - \sum_{k\in K_i} \frac{p_{\bar{k}}(t)\mu_k}{p_{k}(t)} \right), \tag{4.98}$$

where $\bar{K}_i$ is the set of components that are down in system state i, and K_i is the set of components that are up in system state i. This means we can use (4.98) to convert failure probability of the system to frequency of success.

Alternatively, we can use (4.86) to calculate frequency of success from success probability as follows:

$$Fr(t)_{Y\to} = \sum_{k=1}^{n} \left(\frac{\sum_{i\in(S\setminus Y)_k} p_i(t)}{p_k(t)} - \frac{\sum_{i\in(S\setminus Y)_{\bar{k}}} p_i(t)}{p_{\bar{k}}(t)} \right) \times p_{\bar{k}}(t)\mu_k$$

$$= \sum_{k=1}^{n} \sum_{i\in(S\setminus Y)_k} \left(p_i(t)\frac{p_{\bar{k}}(t)\mu_k}{p_k(t)} \right) - \sum_{k=1}^{n} \sum_{i\in(S\setminus Y)_{\bar{k}}} p_i(t)\mu_k.$$

We can then write it compactly as follows:

$$Fr(t)_{Y\to} = \sum_{i\in S\setminus Y} p_i(t) \left(\sum_{k\in K_i} \frac{p_{\bar{k}}(t)\mu_k}{p_k(t)} - \sum_{k\in\bar{K}_i} \mu_k \right), \tag{4.99}$$

where $\bar{K}_i$ is the set of components that are down in system state i, and K_i is the set of components that are up in system state i. This means that the frequency can be calculated from weighted summation of probability of a success state i with $\frac{p_{\bar{k}}(t)\mu_k}{p_k(t)}$ when the component k is up in state i and subtracting with the repair rates of all components k that are down in state i. Equation (4.99) is an alternative formula to calculate frequency of encountering success states using success probability.

Similarly, we can rewrite (4.85) as follows:

$$Fr(t)_{\to Y} = \sum_{k=1}^{n} \left(\frac{\sum_{i\in(S\setminus Y)_k} p_i(t)}{p_k(t)} - \frac{\sum_{i\in(S\setminus Y)_{\bar{k}}} p_i(t)}{p_{\bar{k}(t)}} \right) \times p_k(t)\lambda_k$$

$$= \sum_{k=1}^{n} \sum_{i\in(S\setminus Y)_k} p_i(t)\lambda_k - \sum_{k=1}^{n} \sum_{i\in(S\setminus Y)_{\bar{k}}} \left(p_i(t)\frac{p_k(t)\lambda_k}{p_{\bar{k}(t)}} \right).$$

This means that we can calculate the frequency of entering failure states from weighted summation of probability of success state i, $p_i(t)$ with some terms. Note that the first term denotes the summation of failure rates of all components that are up in state i, and the second term denote the summation of all components that are down in state i.

Since $S \setminus Y = (S \setminus Y)_k \cup (S \setminus Y)_{\bar{k}}$, we can write (4.85) compactly as follows:

$$Fr(t)_{\to Y} = \sum_{i\in S\setminus Y} p_i(t) \left(\sum_{k\in K_i} \lambda_k - \sum_{k\in\bar{K}_i} \frac{p_k(t)\lambda_k}{p_{\bar{k}(t)}} \right), \tag{4.100}$$

where K_i is the set of components that are up in system state i, and $\bar{K}_i$ is the set of components that are down in system state i. This means we can use (4.100) to convert failure probability of the system to frequency of success.

Alternatively, we can use (4.86) to calculate frequency of failure from system probability of failure as follows:

$$
Fr(t)_{\to Y} = \sum_{k=1}^{n} \left(\frac{\sum_{i \in Y_{\bar{k}}} p_i(t)}{p_{\bar{k}}(t)} - \frac{\sum_{i \in Y_k} p_i(t)}{p_k(t)} \right) \times p_k(t)\lambda_k
$$

$$
= \sum_{k=1}^{n} \sum_{i \in Y_{\bar{k}}} \left(p_i(t) \frac{p_k(t)\lambda_k}{p_{\bar{k}}(t)} \right) - \sum_{k=1}^{n} \sum_{i \in Y_k} p_i(t)\lambda_k.
$$

We can then write it compactly as follows:

$$
Fr(t)_{\to Y} = \sum_{i \in Y} p_i(t) \left(\sum_{k \in \bar{K}_i} \frac{p_k(t)\lambda_k}{p_{\bar{k}}(t)} - \sum_{k \in K_i} \lambda_k \right), \tag{4.101}
$$

where $\bar{K}_i$ is the set of components that are down in system state i, and K_i is the set of components that are up in system state i. This means that the frequency of failure can be calculated from weighted summation of probability of a failure state i with $\frac{p_k(t)\lambda_k}{p_{\bar{k}}(t)}$ when the component k is down in state i and subtracting with the failure rates of all components k that are up in state i.

This means that we can now find time-specific frequency of success using failure probability of the system from (4.98) and time-specific frequency of failure using success probability of the system from (4.100). Alternatively, we can also convert the success (or failure) probability of the system to frequency of success (or failure) from (4.99) and (4.101), respectively, as will be shown in Example 4.11.

Example 4.11 Consider a two-unit redundant system as shown in Figure 4.20. Assuming an independent and coherent system, we can describe success probability and failure probability of the system as follows:

$$
p_s(t) = p_1(t)p_2(t) + p_1(t)p_{\bar{2}}(t) + p_{\bar{1}}(t)p_2(t)
$$
$$
p_f(t) = p_{\bar{1}}(t)p_{\bar{2}}(t).
$$

Let us first find the frequency of encountering success states using (4.98):

$$
f_s(t) = p_{\bar{1}}(t)p_{\bar{2}}(t)(\mu_1 + \mu_2).
$$

Figure 4.20 A two-unit redundant system.

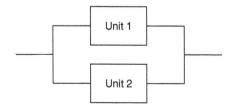

We can also find the same index using (4.99) as follows:

$$f_s(t) = p_1(t)p_2(t)\left(\frac{\mu_1 p_{\bar{1}}(t)}{p_1(t)} + \frac{\mu_2 p_{\bar{2}}(t)}{p_2(t)}\right) + p_1(t)p_{\bar{2}}(t)\left(\frac{\mu_1 p_{\bar{1}}(t)}{p_1(t)} - \mu_2\right)$$

$$+ p_{\bar{1}}(t)p_2(t)\left(\frac{\mu_2 p_{\bar{1}}(t)}{p_2(t)} - \mu_1\right)$$

$$= p_{\bar{1}}(t)p_{\bar{2}}(t)(\mu_1 + \mu_2),$$

which is identical to what we obtained using (4.98).

Similarly, we can find the frequency of failure using (4.100):

$$f_f(t) = p_1(t)p_2(t)(\lambda_1 + \lambda_2) + p_1(t)p_{\bar{2}}(t)\left(\lambda_1 - \frac{\lambda_2 p_2(t)}{p_{\bar{2}}(t)}\right)$$

$$+ p_{\bar{1}}(t)p_2(t)\left(\lambda_2 - \frac{\lambda_1 p_1(t)}{p_{\bar{1}}(t)}\right)$$

$$= p_1(t)p_{\bar{2}}(t)\lambda_1 + p_{\bar{1}}(t)p_2(t)\lambda_2.$$

We can obtain the same expression using the failure probability of the system in (4.101) as follows:

$$f_f(t) = p_{\bar{1}}(t)p_{\bar{2}}(t)\left(\frac{\lambda_1 p_1(t)}{p_{\bar{1}}(t)} + \frac{\lambda_2 p_2(t)}{p_{\bar{2}}(t)}\right)$$

$$= p_1(t)p_{\bar{2}}(t)\lambda_1 + p_{\bar{1}}(t)p_2(t)\lambda_2,$$

which is identical to what we obtained using (4.100).

We can now derive the conversion rules for steady state domain as follows:

Rule 1 Convert probability of success to frequency of success using (4.98):

$$Fr_{Y\rightarrow} = \sum_{i\in S\backslash Y} p_i\left(\sum_{k\in K_i}\frac{p_{\bar{k}}\mu_k}{p_k} - \sum_{k\in \bar{K}_i}\mu_k\right). \tag{4.102}$$

Rule 2 Convert probability of failure to frequency of success using (4.99):

$$Fr_{Y\rightarrow} = \sum_{i\in Y} p_i\left(\sum_{k\in \bar{K}_i}\mu_k - \sum_{k\in K_i}\frac{p_{\bar{k}}\mu_k}{p_k}\right). \tag{4.103}$$

Rule 3 Convert probability of success to frequency of failure using (4.100):

$$Fr_{\rightarrow Y} = \sum_{i\in S\backslash Y} p_i\left(\sum_{k\in K_i}\lambda_k - \sum_{k\in \bar{K}_i}\frac{p_k\lambda_k}{p_{\bar{k}}}\right). \tag{4.104}$$

Rule 4 Convert probability of failure to frequency of failure using (4.101):

$$Fr_{\rightarrow Y} = \sum_{i \in Y} p_i \left(\sum_{k \in \bar{K}_i} \frac{p_k \lambda_k}{p_{\bar{k}}} - \sum_{k \in K_i} \lambda_k \right), \tag{4.105}$$

where

Y Set of failure states
$S \setminus Y$ Set of success states
p_i Probability of state i
K_i Set of components that is up in system state i
$\bar{K}_i$ Set of components that is down in system state i
p_k Probability of component k in up state
$p_{\bar{k}}$ Probability of component k in down state
μ_k Repair rate of component k
λ_k Failure rate of component k

Exercises

4.1 Consider a system of two components. Each component can assume two statuses: failure or success. A failure of either component will cause a system to fail. Each component has a failure rate of $\lambda_1 = \lambda_2 = 0.1$ per year and a repair rate of of $\mu_1 = \mu_2 = 10$ per year. In a rare occasion, both components at success state will fail at the same time at a rate of $\lambda_c = 0.01$ per year. (We will later call this mode of failure *common mode failure* and the rate at which both components in working condition fails together *common mode failure rate* in Chapters 5 and 9). Show that this system is *not frequency balanced*. Hint: first find the transition-rate matrix and calculate steady state probability. Draw any boundary between two states and calculate two frequencies.

4.2 Consider a system of three components shown in Figure 4.10. Find the steady state probability using a frequency balance technique. Verify your result by using the transition-rate matrix.

5

Analytical Methods in Reliability Analysis

5.1 Introduction

The main objective of system reliability measures or indices is to quantify the behavior of the system related to its failures. Basic reliability measures are probability of failure, frequency of failure, mean cycle time, mean down time and mean up time. In fact, once the probability and frequency of failure are found, the rest can be simply calculated from those two. Over the course of time many specific methods for reliability analysis have been developed and widely used. These methods are typically categorized into analytical methods and methods based on the Monte Carlo simulation.

This chapter describes some of the commonly used analytical methods in power system reliability evaluation. The following methods will be described in detail:

1. State space method using Markov processes.
2. Network reduction method.
3. Conditional probability method.
4. Cut-set or Tie-set method.

5.2 State Space Approach

When the system is described as a Markov process, we can represent its stochastic behavior using a state-transition diagram as mentioned in Chapter 3. The system behavior can be constructed from the random behavior of each component in the system, which is typically presented as a component's state-transition diagram. This means that we can build the system state space comprising all possible combinations of states of components comprising the system. The system state-transition diagram shows all possible states that the system can assume and explains how the system evolves to other states via a transition rate between states.

Electric Power Grid Reliability Evaluation: Models and Methods, First Edition. Chanan Singh,
Panida Jirutitijaroen, and Joydeep Mitra.

The state space approach is used to calculate reliability indices when all system states are described by a state-transition diagram. Reliability indices such as probability of failure, frequency of failure and duration of failure can be easily calculated once we know all the information about system state space. The state space approach can be considered the most direct approach to calculate reliability indices as all possible system states are enumerated. It is quite flexible and can be used to solve many problems of interest in power systems.

Once the system states are enumerated, we can classify each system state into one of the two categories: failure or success. The probability of failure is found from the summation of state probabilities of all failure states. When state probabilities are known, frequency of failure can be calculated directly by drawing a boundary between failure and success states. Duration indices can be calculated from probability and frequency indices.

The state space approach consists of the following steps:

Step 1 Identify all possible states.
Step 2 Determine transition rates between states.
Step 3 Calculate state probabilities.
Step 4 Calculate reliability indices

When the components are assumed to fail and be repaired independently, i.e., they are *independent*, then step 3 can be omitted, as the state probabilities can be found simply by multiplying the probabilities of component states. Otherwise, the state probabilities can be calculated using the knowledge of transition rate matrix.

In the following section, we first explain how to construct the system state-transition diagram using the component's state-transition diagram. We later describe the construction of transition-rate matrix in Section 5.2.2. In Section 5.2.3 the state probability calculation, as explained in Chapter 3, will be repeated again here in a matrix formulation. The calculation of reliability indices is given in Section 5.2.4.

5.2.1 System State-Transition Diagram

The system state-transition diagram is constructed using each individual component's state-transition diagram. It provides visual representation of all system states and the manner in which system can make transition between states. Basically, system states are enumerated from overall system activity as a result of changes in individual component states as well as conditions and constraints in which the overall system works. The transition rates from one system state to others are determined from the change of component status and other factors, such as dependencies between components and weather changes. The system state diagram provides information on how the system moves from one state to others via component transition rates.

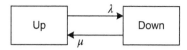

Figure 5.1 A two-state component.

When the system state-transition diagram includes all possible combinations of component states, each system state can be reached by changing one component state at a time, and all the component transitions are realized, we say the components are *independent*. This implies that the failure and repair of one component does not interfere with the failure and repair of other components and that they follow their stochastic behavior independently.

For a system comprising of n independent components and each component having m_k states, the entire system state space contains all possible states resulting in $\prod_{k=1}^{n} m_k$ states. The following example illustrates the state-transition diagram of a two-independent-component system where each component has two states.

Example 5.1 For a two-component system, each component state-transition diagram is shown in Figure 5.1.

The state-transition diagram of this system is shown in Figure 5.2.

In some cases, although the system contains all possible combinations of component states, some transitions of certain components in some system states may not be realized. For example, a system may have a sequence of component repair in a fixed sequential order, called *restricted repair*. This means that when two or more components fail, we can only repair a particular component before others. Consequently, the repair rates of some components may not be realized in some states. An example of state transition diagram of this system is shown in Example 5.2.

Example 5.2 The system consists of two components whose state-transition diagram is shown in Figure 5.1. Due to a maintenance budget limitation there is a limited repair crew, and they can only fix one component at a time when it fails. In addition, the system has a priority to always bring back component

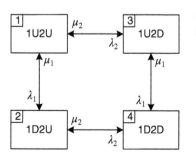

Figure 5.2 A two-independent-component system.

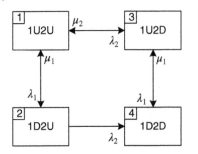

Figure 5.3 A two-component system in the case of restricted repair.

one into operation before repairing component two. When both components fail, component one has priority to be fixed first. The state-transition diagram of this system is shown in Figure 5.3.

We can see from this example that the repair rate of component two can never be realized when both components fail. The system will always repair component one first and transit from state four back to state three. As a result, there will be no transition from state four to state two as would be possible if the repair processes of the two components were independent.

It is also possible that the system state space may not contain all possible combinations of component states. A typical example of such system is when the failure of one component causes the system to fail and further component failures will not occur. This means that the system state where two or more component fail will never exist. The failures of components of this system are called *dependent* failures. A state-transition diagram of such a system is depicted in Figure 5.4. We will revisit this system again in Section 5.3.

For some rare events, the system may move from one state to another state when two or more components change their status in one transition. When the system can access failure states by changing status of two or more components at the same time, we call the failure in this situation *common-mode failure*. The classic example is when two transmission lines are hit by lightning and cause both lines to breakdown at the same time. An example of the state-transition diagram for common-mode failure is given in Example 5.3.

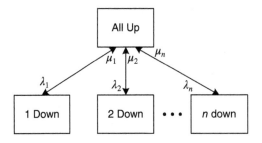

Figure 5.4 A state-transition diagram of dependent-failure system.

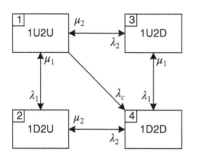

Figure 5.5 A system state-transition diagram for common-mode failure.

Example 5.3 For a two-component system, each component state-transition diagram is given in Figure 5.1. In less-frequent cases, both components can fail at the same time with a transition rate of λ_c. The state-transition diagram of this system is shown in Figure 5.5.

As we can see from this section, the system behavior governs the states and transition rate between states in the state-transition diagram. Once the system state-transition diagram is constructed, we can now derive a transition-rate matrix from this diagram, as described in the following section.

5.2.2 Transition-Rate Matrix

The details of constructing the transition-rate matrix are explained in the following steps:

Step 1 Identify all possible system states resulting from failure and repair of components according to the system's behavior, as explained in Section 5.2.1.
Step 2 The interstate transition rates are next determined and are shown in a state-transition diagram.
Step 3 The element of transition-rate matrix **R** are such that

$$R_{ij} = \begin{cases} \lambda_{ij} & i \neq j \\ -\sum_j \lambda_{ij} & i = j \end{cases}, \tag{5.1}$$

where λ_{ij} is transition-rate from state i to state j.

We illustrate the process of constructing a transition rate matrix by the following examples.

Example 5.4 For a system consisting of two independent components, the state-transition diagram is shown in Figure 5.2. We can observe that, comparing states one and two, the difference is failure of component one, so the transition rate from state one to two is the failure rate of component one. We can observe similar transitions for all other states. Then, we can write the transition rate matrix as follows:

$$
\mathbf{R} = \begin{bmatrix}
-\lambda_1 - \lambda_2 & \lambda_1 & \lambda_2 & 0 \\
\mu_1 & -\mu_1 - \lambda_2 & 0 & \lambda_2 \\
\mu_2 & 0 & -\mu_2 - \lambda_1 & \lambda_1 \\
0 & \mu_2 & \mu_1 & -\mu_1 - \mu_2
\end{bmatrix},
$$

where λ_k, μ_k are failure and repair rates of component k.

Example 5.5 Considering the system in Example 5.2, with restricted repair. As shown in Figure 5.3, we can write the transition rate matrix as follows:

$$
\mathbf{R} = \begin{bmatrix}
-\lambda_1 - \lambda_2 & \lambda_1 & \lambda_2 & 0 \\
\mu_1 & -\mu_1 - \lambda_2 & 0 & \lambda_2 \\
\mu_2 & 0 & -\mu_2 - \lambda_1 & \lambda_1 \\
0 & 0 & \mu_1 & -\mu_1
\end{bmatrix},
$$

where λ_k, μ_k are failure and repair rates of component k.

Example 5.6 For a two-component system with dependent failure, the state-transition diagram is given in Figure 5.6.

We can observe that the system consists of only three states. The system can move from state one to state two as a result of failure of component one. Likewise, the system moves from state one to state three as a result of failure of component two. There is no transition between states two and three. We can write the transition rate matrix as follows:

$$
\mathbf{R} = \begin{bmatrix}
-\lambda_1 - \lambda_2 & \lambda_1 & \lambda_2 \\
\mu_1 & -\mu_1 & 0 \\
\mu_2 & 0 & -\mu_2
\end{bmatrix},
$$

where λ_k, μ_k are failure and repair rates of component k.

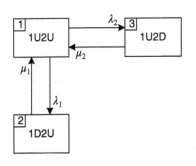

Figure 5.6 A system state-transition diagram for a two-component system with dependent failure.

Example 5.7   For a two-component system with common-mode failure, the state-transition diagram is given by Figure 5.5. In this case, the state space has four states similar to the case shown in Example 5.4. The difference is that in this example, there is one extra transition rate from state one to state four. This transition rate describes the system behavior that both components can fail at the same time. We can write the transition rate matrix as follows:

$$
\mathbf{R} = \begin{bmatrix}
-\lambda_1 - \lambda_2 - \lambda_c & \lambda_1 & \lambda_2 & \lambda_c \\
\mu_1 & -\mu_1 - \lambda_2 & 0 & \lambda_2 \\
\mu_2 & 0 & -\mu_2 - \lambda_1 & \lambda_1 \\
0 & \mu_2 & \mu_1 & -\mu_1 - \mu_2
\end{bmatrix},
$$

where λ_k, μ_k are failure and repair rates of component k, and λ_c is a common-mode failure rate of this system.

The transition-rate matrix can be considered as a formal mathematical representation of a state transition diagram. Since the process is Markovian, we can find the equilibrium state probabilities using the transition-rate matrix in the following section.

5.2.3   Calculation of State Probability

We have previously described the process to calculate equilibrium state probabilities in the Markov processes in Chapter 3 and the frequency-based approach in Chapter 4. Generally, we can formulate the calculation in a matrix form and obtain the steady state probabilities by solving (5.2).
From $\mathbf{pR} = 0$ and $\sum_{\forall i} p_i = 1$, we can write

$$
\mathbf{p} = \mathbf{V}_R \cdot \mathbf{R}'^{-1}, \tag{5.2}
$$

where

$\mathbf{R}'$ Matrix obtained from a transition-rate matrix, $\mathbf{R}$, by replacing the elements of an arbitrarily selected column j with 1's.
$\mathbf{p}$ Steady state probability vector, $\mathbf{p} = [p_1, p_2, \ldots, p_i, \ldots]$; p_i is the steady state probability of the system being in state i.
$\mathbf{V}_R$ Row vector with the jth element equal to 1 and other elements set to 0.

We can use (5.2) to calculate equilibrium state probabilities as shown in the examples below.

Example 5.8   Consider the same system shown in Example 5.4; we can write (5.2) for this system when replacing column 1 values in the transition-rate matrix with 1's as follows:

$$
[p_1 \quad p_2 \quad p_3 \quad p_4] = [1 \quad 0 \quad 0 \quad 0]
\begin{bmatrix}
1 & \lambda_1 & \lambda_2 & 0 \\
1 & -\mu_1 - \lambda_2 & 0 & \lambda_2 \\
1 & 0 & -\mu_2 - \lambda_1 & \lambda_1 \\
1 & \mu_2 & \mu_1 & -\mu_1 - \mu_2
\end{bmatrix}^{-1}.
$$

We then have

$$
p_1 = \frac{\mu_1 \mu_2}{(\lambda_1 + \mu_1)(\lambda_2 + \mu_2)}
$$

$$
p_2 = \frac{\lambda_1 \mu_2}{(\lambda_1 + \mu_1)(\lambda_2 + \mu_2)}
$$

$$
p_3 = \frac{\mu_1 \lambda_2}{(\lambda_1 + \mu_1)(\lambda_2 + \mu_2)}
$$

$$
p_4 = \frac{\lambda_1 \lambda_2}{(\lambda_1 + \mu_1)(\lambda_2 + \mu_2)}.
$$

It should be observed from this example that when the components are independent, the state probabilities are just the product of component probabilities. For example, in system state one the components one and two are working. Then, the equilibrium probability of state one can be found by multiplying the probability that component one is working by the probability that component two is working.

Example 5.9   Consider the system shown in Example 5.5; we can write (5.2) for this system when replacing column 1 values in the transition rate matrix with 1's as follows:

$$
[p_1 \quad p_2 \quad p_3 \quad p_4] = [1 \quad 0 \quad 0 \quad 0]
\begin{bmatrix}
1 & \lambda_1 & \lambda_2 & 0 \\
1 & -\mu_1 - \lambda_2 & 0 & \lambda_2 \\
1 & 0 & -\mu_2 - \lambda_1 & \lambda_1 \\
1 & 0 & \mu_1 & -\mu_1
\end{bmatrix}^{-1}.
$$

We then have

$$
p_1 = \frac{\mu_1 \mu_2(\lambda_2 + \mu_1)}{\lambda_1 \lambda_2(\lambda_1 + \lambda_2 + 2\mu_1 + \mu_2) + \mu_1 \mu_2(\lambda_1 + \lambda_2 + \mu_1) + \lambda_2 \mu_1(\lambda_2 + \mu_1)}
$$

$$
p_2 = \frac{\lambda_1 \mu_1 \mu_2}{\lambda_1 \lambda_2(\lambda_1 + \lambda_2 + 2\mu_1 + \mu_2) + \mu_1 \mu_2(\lambda_1 + \lambda_2 + \mu_1) + \lambda_2 \mu_1(\lambda_2 + \mu_1)}
$$

$$
p_3 = \frac{\lambda_2 \mu_1(\lambda_1 + \lambda_2 + \mu_1)}{\lambda_1 \lambda_2(\lambda_1 + \lambda_2 + 2\mu_1 + \mu_2) + \mu_1 \mu_2(\lambda_1 + \lambda_2 + \mu_1) + \lambda_2 \mu_1(\lambda_2 + \mu_1)}
$$

$$
p_4 = \frac{\lambda_1 \lambda_2(\lambda_1 + \lambda_2 + \mu_1 + \mu_2)}{\lambda_1 \lambda_2(\lambda_1 + \lambda_2 + 2\mu_1 + \mu_2) + \mu_1 \mu_2(\lambda_1 + \lambda_2 + \mu_1) + \lambda_2 \mu_1(\lambda_2 + \mu_1)}.
$$

Example 5.10   Consider a two-component system with dependent failure; the state-transition diagram is given in Figure 5.6. The transition-rate matrix is found in Example 5.6. We can use (5.2) to calculate the equilibrium state probability as follows:

$$[p_1 \quad p_2 \quad p_3] = [1 \quad 0 \quad 0] \begin{bmatrix} 1 & \lambda_1 & \lambda_2 \\ 1 & -\mu_1 & 0 \\ 1 & 0 & -\mu_2 \end{bmatrix}^{-1}.$$

We then have

$$p_1 = \frac{\mu_1\mu_2}{\mu_1\mu_2 + \lambda_1\mu_2 + \lambda_2\mu_1}$$

$$p_2 = \frac{\lambda_1\mu_2}{\mu_1\mu_2 + \lambda_1\mu_2 + \lambda_2\mu_1}$$

$$p_3 = \frac{\lambda_2\mu_1}{\mu_1\mu_2 + \lambda_1\mu_2 + \lambda_2\mu_1}.$$

After the state probabilities are computed, we need to identify the failure states in the system. Reliability indices can then be calculated accordingly.

5.2.4   Calculation of Reliability Indices

Once the state probabilities have been found, the reliability indices can be easily found as explained below.

Probability of system failure
The probability of system failure is easily found by summing up the probabilities of failure states, as given by (1.1).

Frequency of system failure
This is the expected number of failures per unit time. From Chapter 4, this can be easily obtained by finding the expected transitions entering the boundary of subset Y of failure states and is given by (4.17). Alternatively, the frequency of failure can be calculated using (4.18) due to frequency balance at the steady state.

It should be noted that although (4.17) or (4.18) can be employed for calculating the frequency of failure index, for relatively large number of states, these are not convenient to apply. We present a matrix approach to calculate reliability indices in the following.

Matrix approach to calculate frequency of system failure
For larger systems, a matrix approach can be used as a systematic approach to calculate the frequency of failure.

Consider a transition-rate matrix, $\mathbf{R}$; each off-diagonal elements represents a transition rate from state i to state j; let $\bar{\mathbf{R}}$ be the transition-rate matrix whose diagonal elements are set to zero:

$$
\bar{\mathbf{R}} = \begin{bmatrix}
0 & \lambda_{12} & \lambda_{13} & \cdots & \lambda_{1n} \\
\lambda_{21} & 0 & \lambda_{23} & \cdots & \lambda_{2n} \\
\lambda_{31} & \lambda_{32} & 0 & \cdots & \lambda_{3n} \\
\vdots & \vdots & \vdots & \vdots & \vdots \\
\lambda_{n1} & \lambda_{n2} & \cdots & \lambda_{n,n-1} & 0
\end{bmatrix}. \tag{5.3}
$$

When we multiply the state probability with this modified transition-rate matrix, we have a row vector whose element jth is given by (5.4):

$$
(\mathbf{p}\bar{\mathbf{R}})_j = \sum_{i \in S \setminus \{j\}} p_i \lambda_{ij}. \tag{5.4}
$$

Recall from Chapter 4, equation 4.9, that (5.4) is the frequency of encountering state j from all other states i in the state space. We can use (5.4) to find the frequency of encountering state j from success states by modifying the state probability vector to only account for the success states as follows.

Let $\bar{\mathbf{p}}$ be a modified state probability derived from steady state probability by replacing the element i with zero, $\forall i \in Y$, failure states. When we multiply $\bar{\mathbf{p}}$ with $\bar{\mathbf{R}}$, we have a row vector whose element jth is now given by (5.5):

$$
(\bar{\mathbf{p}}\bar{\mathbf{R}})_j = \sum_{i \in S \setminus Y} p_i \lambda_{ij}, \tag{5.5}
$$

which is the frequency from success states $i \in S \setminus Y$ to any states j in the state space. The frequency from success to failure states is found by adding all the elements associated with failure states $j \in Y$ as follows:

$$
f_f = (\bar{\mathbf{p}}\bar{\mathbf{R}}) \cdot \mathbf{1}_{(j \in Y)}, \tag{5.6}
$$

where

$\mathbf{1}_{(j \in Y)}$ A $n \times 1$ vector of indicator function $\mathbf{1}_{(i \in Y)}$, which equals one for the failure states in Y, and zero otherwise.

$\bar{\mathbf{R}}$ Transition-rate matrix with its diagonal elements set to zero.

$\bar{\mathbf{p}}$ Modified steady state probability by replacing the element i associated with failure states with zero.

We can also find frequency of encountering success states in a matrix form. Similar to (5.5), the frequency from failure states to any other states can be found as follows:

$$
((\mathbf{p} - \bar{\mathbf{p}})\bar{\mathbf{R}})_j = \sum_{i \in Y} p_i \lambda_{ij}. \tag{5.7}
$$

Note that $\mathbf{p} - \bar{\mathbf{p}}$ yields the vector of steady state probabilities, replacing its element i associated with success states with zero. The frequency to success states is found by adding all the elements associated with success states $j \in S \setminus Y$ as follows:

$$f_s = ((\mathbf{p} - \bar{\mathbf{p}})\bar{\mathbf{R}}) \cdot \mathbf{1}_{(j \in S \setminus Y)}, \tag{5.8}$$

where f_s is frequency of success and $\mathbf{1}_{(j \in S \setminus Y)}]$ is a $n \times 1$ vector of indicator function $\mathbf{1}_{(i \in S \setminus Y)}$, which equals one for the success states in $S \setminus Y$, and zero otherwise.

In the steady state according to the frequency-balance concept in Chapter 4, the frequency of failure is the same as frequency of success and both (5.6) and (5.8) can be used to calculate this frequency of failure. We illustrate the matrix approach to calculate both frequency of failure and frequency of success in the following example.

Example 5.11   Consider the same system as shown in Example 5.4. Assume that the system fails when either one of the component fails, i.e., the failure states are state two, three and four. We can find $\bar{\mathbf{R}}$ and $\bar{\mathbf{p}}$ as follows:

$$\bar{\mathbf{R}} = \begin{bmatrix} 0 & \lambda_1 & \lambda_2 & 0 \\ \mu_1 & 0 & 0 & \lambda_2 \\ \mu_2 & 0 & 0 & \lambda_1 \\ 0 & \mu_2 & \mu_1 & 0 \end{bmatrix}, \bar{\mathbf{p}} = [p_1 \quad 0 \quad 0 \quad 0].$$

Using (5.6), we have

$$\mathbf{1}_{(j \in Y)} = \begin{bmatrix} 0 \\ 1 \\ 1 \\ 1 \end{bmatrix}.$$

Then,

$$\begin{aligned} f_f &= (\bar{\mathbf{p}}\bar{\mathbf{R}}) \cdot \mathbf{1}_{(j \in Y)} \\ &= [0 \quad p_1 \lambda_1 \quad p_1 \lambda_2 \quad 0] \cdot \mathbf{1}_{(j \in Y)} \\ &= p_1 (\lambda_1 + \lambda_2), \end{aligned}$$

which is the frequency of encountering failure states from success states and can be easily verified by the state-transition diagram.

Alternatively, we can use (5.8) as follows:

$$(\mathbf{p} - \bar{\mathbf{p}}) = [0 \quad p_2 \quad p_3 \quad p_4], \mathbf{1}_{(j \in S \setminus Y)} = \begin{bmatrix} 1 \\ 0 \\ 0 \\ 0 \end{bmatrix}.$$

Then,

$$f_s = ((\mathbf{p} - \bar{\mathbf{p}})\bar{\mathbf{R}}) \cdot \mathbf{1}_{(j \in S \setminus Y)}$$
$$= [p_2\mu_1 + p_3\mu_2 \quad p_4\mu_2 \quad p_4\mu_1 \quad p_2\lambda_2 + p_3\lambda_1] \cdot \mathbf{1}_{(j \in S \setminus Y)}$$
$$= p_2\mu_1 + p_3\mu_2,$$

which is the frequency of encountering success states from failure states and can be verified by the state-transition diagram.

Once the probability and frequency of failure are found, other indices, such as mean cycle time, mean down time, and mean up time can be simply calculated using (1.2), (1.3) and (1.4), respectively.

Although the state space approach is the most direct method to compute reliability indices, we can observe that the main drawback of this approach is that it requires all possible states to be enumerated in the state space diagram. When the system contains numerous components, a transition rate diagram or matrix of the whole system may be impractical to construct for a large system. We can overcome this drawback by decomposing the system into subsystems the where state-transition diagram for each subsystem can be easily developed.

5.2.5   Sequential Model Building in the State Space Approach

Sequential model building refers to a process where the system is decomposed into subsystems. The models of subsystems are developed and then combined to form the model of the whole system. During this process some similar states in the subsystems are combined, thus reducing the number of states. Recall from Chapter 4 that we can combine states and obtain equivalent transition rates between the subsets of merged states using (4.57), which is given again below as (5.9):

$$\lambda_{XY} = \frac{\sum_{i \in X} \sum_{j \in Y} p_i \lambda_{ij}}{\sum_{i \in X} p_i}, \tag{5.9}$$

where λ_{XY} is the equivalent transition rate from subset X to subset Y, $X, Y \subset S$.

The system state-transition diagram is then constructed by combining state-transition diagrams of each subsystem. Sequential model building is illustrated by the following example, which is a simplified model of a power system.

Example 5.12   A power system shown in Figure 5.7 consists of three independent subsystems, namely, generation, transmission and load. Components in all subsystems are also assumed to be independent. We focus on the events that the system is not able to satisfy the load, i.e., loss of load events. We are interested in calculating the power system reliability indices which are loss of

Generators   Transmission   Load
                         lines

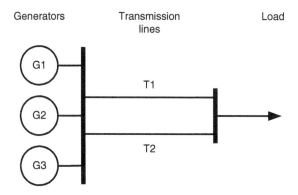

Figure 5.7 A case study of a power system.

load probability, frequency of loss of load and mean duration of loss of load. The data of each subsystem is given below:

- Generation subsystem comprises three generators; each generator has full capacity of 50 MW or 0 MW when failed. Failure rate of each generator is 0.01 per day, and mean repair time is 12 hours.
- Transmission line subsystem contains two transmission lines connected in parallel. The failure rate of each transmission line is 10 per year, and the mean down time is 8 hours. Each transmission line has a capacity of 100 MW or 0 MW when failed.
- The load fluctuates between two states, 150 MW and 50 MW, with mean duration in each state of 8 hours and 16 hours, respectively.

We can see that if we apply the state space approach straightforwardly, we need to enumerate all possible system states in the state space diagram. In this case, the total number of components in this system is 6 (3 generators, 2 transmission lines, and load) and the number of states is 2^6, which is 64. If we follow the steps to enumerate all possible system states, we need to draw a system state-transition diagram for all the 64 states. Although in this case application of direct state space is possible, we will use this system to illustrate the process of sequential model building for larger systems.

We now apply the sequential model building technique by considering each subsystem separately and combining some states within the subsystems. Then, we can construct the system state space using the simplified model of each subsystem.

Generation subsystem

The subsystem consists of three generators. The state-transition diagram of each generator is a two-state Markov model, which is shown in Figure 5.8.

The failure rate of each generator is 0.01 per day, and we know from Chapter 3 that the repair rate is just a reciprocal of the average time that the

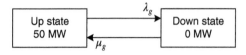

Figure 5.8 A state-transition diagram of a generator.

component is in down state. Thus, we can find the repair rate of a generator to be $\frac{1}{12}$ per hour. To be consistent with all the components in this case study, we present the two rates in the same unit of per year as follows:

$$\lambda_g = 0.01 \text{ per day} = 3.65 \text{ per year}$$

$$\mu_g = \frac{1}{12} \text{ per hour} = 730 \text{ per year}.$$

Recall from Example 3.11 that the probability of a two-state component model can be found from its failure rate λ and repair rate μ and is given again below:

$$p_u = \frac{\mu}{\mu + \lambda}$$

$$p_d = \frac{\lambda}{\mu + \lambda},$$

where p_u and p_d are the probability of a component being in up and down state, respectively.

We can draw the state transition diagram of the generation subsystem of three independent generators as seen in Figure 5.9.

Let p_{gi} be the probability of the generation subsystem state $i, i = \{1, 2, \ldots, 8\}$. Each generator is identical, with the same failure and repair rate. The probabilities of each generator being in up and down state are $\frac{\mu_g}{\mu_g + \lambda_g}$ and $\frac{\lambda_g}{\mu_g + \lambda_g}$, respectively. We can use multiplication rule to compute the state probability as follows:

$$p_{g1} = (\frac{\mu_g}{\mu_g + \lambda_g})^3 = 0.995^3 = 9.851 \times 10^{-1}$$

$$p_{g2} = p_{g3} = p_{g4} = (\frac{\mu_g}{\mu_g + \lambda_g})^2(\frac{\lambda_g}{\mu_g + \lambda_g}) = 0.995^2 \times 0.005 = 4.950 \times 10^{-3}$$

$$p_{g5} = p_{g6} = p_{g7} = (\frac{\mu_g}{\mu_g + \lambda_g})(\frac{\lambda_g}{\mu_g + \lambda_g})^2 = 0.995 \times 0.005^2 = 2.488 \times 10^{-5}$$

$$p_{g8} = (\frac{\lambda_g}{\mu_g + \lambda_g})^3 = 0.005^3 = 1.250 \times 10^{-7}.$$

Notice that states 2, 3 and 4 of Figure 5.9 represent the same identical generating capacity of 100 MW. Likewise, states 5, 6 and 7 have the same generating

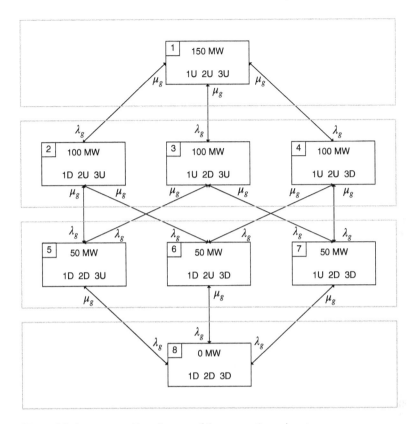

Figure 5.9 A state-transition diagram of the generation subsystem.

capacity of 50 MW. We can merge these states to produce the equivalent state-transition diagram, as shown in Figure 5.10.

The number of system states in this generation subsystem reduces from eight states to four states. Denoting the equivalent transition-rate by λ, the equivalent transition rates can be found as follows. All the units are in per year:

$$\lambda_{g150 \to g100} = \frac{p_{g1}(\lambda_g + \lambda_g + \lambda_g)}{p_{g1}} = 3\lambda_g = 10.95$$

$$\lambda_{g100 \to g50} = \frac{p_{g2}(\lambda_g + \lambda_g) + p_{g3}(\lambda_g + \lambda_g) + p_{g4}(\lambda_g + \lambda_g)}{p_{g2} + p_{g3} + p_{g4}} = 2\lambda_g = 7.30$$

$$\lambda_{g50 \to g0} = \frac{p_{g5}(\lambda_g + p_{g6}\lambda_g + p_{g7}\lambda_g)}{p_{g5} + p_{g6} + p_{g7}} = \lambda_g = 3.65$$

$$\lambda_{g0 \to g50} = \frac{p_{g8}(\mu_g + \mu_g + \mu_g)}{p_{g8}} = 3\mu_g = 2190$$

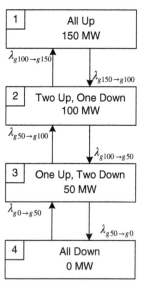

Figure 5.10 The equivalent state-transition diagram of the generation subsystem.

$$\lambda_{g50 \to g100} = \frac{p_{g5}(\mu_g + \mu_g) + p_{g6}(\mu_g + \mu_g) + p_{g7}(\mu_g + \mu_g)}{p_{g5} + p_{g6} + p_{g7}} = 2\mu_g = 1460$$

$$\lambda_{g100 \to g150} = \frac{p_{g2}\mu_g + p_{g3}\mu_g + p_{g4}\mu_g}{p_{g2} + p_{g3} + p_{g4}} = \mu_g = 730.$$

Denote the probability $p_{g150}, p_{g100}, p_{g50}, p_{g0}$ as the probability of the generation subsystem producing the equivalent of 150, 100, 50 and 0 MW, respectively. The probability of being in each state in the equivalent diagram is found as follows:

$$p_{g150} = p_{g1} = 9.851 \times 10^{-1}$$

$$p_{g100} = p_{g2} + p_{g3} + p_{g4} = 1.485 \times 10^{-2}$$

$$p_{g50} = p_{g5} + p_{g6} + p_{g7} = 7.463 \times 10^{-5}$$

$$p_{g0} = p_{g8} = 1.250 \times 10^{-7}.$$

We now have all the parameters of the equivalent generation subsystem, namely, state probabilities and transition rates.

Transmission subsystem
The state transition diagram of each transmission line is shown in Figure 5.11, with the following parameters:

$$\lambda_t = 10 \text{ per year}$$

$$\mu_t = \frac{1}{8} \text{ per hour} = 1095 \text{ per year}.$$

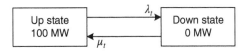

Figure 5.11 The state-transition diagram of a transmission line.

Let p_{ti} be the probability of the transmission subsystem state i, $i = \{1, 2, 3, 4\}$. Each transmission line is identical, with the same failure and repair rate. The probabilities of each transmission line being in up and down state are $\frac{\mu_t}{\mu_t + \lambda_t}$ and $\frac{\lambda_t}{\mu_t + \lambda_t}$, respectively. We can use the multiplication rule to compute the state probability as follows:

$$p_{t1} = (\frac{\mu_t}{\mu_t + \lambda_t})^2 = 9.812 \times 10^{-1}$$

$$p_{t2} = p_{t3} = (\frac{\mu_t}{\mu_t + \lambda_t})(\frac{\lambda_t}{\mu_t + \lambda_t}) = 8.968 \times 10^{-3}$$

$$p_{t4} = (\frac{\lambda_t}{\mu_t + \lambda_t})^2 = 8.190 \times 10^{-5}.$$

When both transmission lines are independent, we have the state-transition diagram of this subsystem as shown in Figure 5.12.

Since we are only interested in the transfer capability of the transmission subsystems, we can combine states with the same capacity to produce an equivalent state-transition diagram for the transmission line subsystem, as shown in Figure 5.13.

The number of system states in generation subsystem reduces from four states to three states. Let the equivalent transition be λ; the equivalent transition rates can be found using (5.9) as follows. All units are in per year:

$$\lambda_{t200 \to t100} = \frac{p_{t1}(\lambda_t + \lambda_t)}{p_{t1}} = 2\lambda_t = 20$$

$$\lambda_{t100 \to t0} = \frac{p_{t2}\lambda_t + p_{t3}\lambda_t}{p_{t2} + p_{t3}} = \lambda_t = 10$$

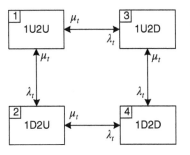

Figure 5.12 The state-transition diagram of two transmission lines.

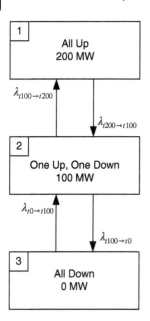

$$\lambda_{t0 \rightarrow t100} = \frac{p_{t4}(\mu_t + \mu_t)}{p_{t4}} = 2\mu_t = 2190$$

$$\lambda_{t100 \rightarrow t200} = \frac{p_{t2}\mu_t + p_{t3}\mu_t}{p_{t2} + p_{t3}} = \mu_t = 1095.$$

Denote the probability $p_{t200}, p_{t100}, p_{t0}$ as the probability of the transmission subsystem capacity of 200, 100 and 0 MW, respectively. The probability of being in each state in the equivalent diagram is found as follows:

$$p_{t200} = p_{t1} = 9.812 \times 10^{-1}$$
$$p_{t100} = p_{t2} + p_{t3} = 1.794 \times 10^{-2}$$
$$p_{t0} = p_{t4} = 8.190 \times 10^{-5}.$$

Combined generation and transmission subsystems

Since the objective of a power system is to deliver the generation capacity to the load, we can now combine the equivalent transition diagram of the generation subsystem with the equivalent transition diagram of the transmission subsystem. The combined capacity is the total capacity delivered at the load point. Recall that the number of states in equivalent generation subsystem is four, and the number of states in equivalent transmission subsystem is three, the number of states in the state transition diagram of this delivered capacity is $3 \times 4 = 12$. The state-transition diagram of this combined subsystem is shown in Figure 5.14. The delivered capacity in each state is found from the $\min\{g, t\}$,

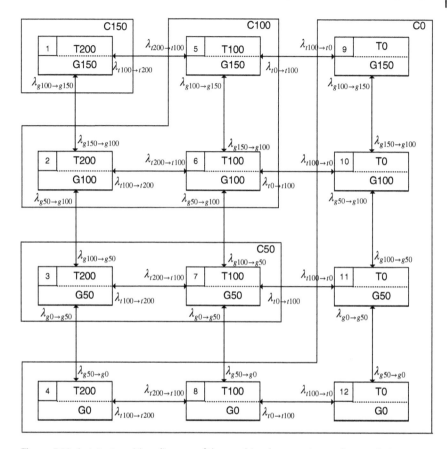

Figure 5.14 A state-transition diagram of the combined generation and transmission subsystem.

the minimum capacity of generation and transmission, and is shown in the diagram.

Let p_{ci} be the probability of the combined generation and transmission state $i, i = \{1, 2, \ldots, 12\}$. The state probabilities are found from the multiplication rule, assuming that all components are independent as follows:

$$p_{c1} = p_{g150}p_{t200} = 9.673 \times 10^{-1}$$
$$p_{c2} = p_{g100}p_{t200} = 1.458 \times 10^{-2}$$
$$p_{c3} = p_{g50}p_{t200} = 7.328 \times 10^{-5}$$
$$p_{c4} = p_{g0}p_{t200} = 1.227 \times 10^{-7}$$
$$p_{c5} = p_{g150}p_{t100} = 1.767 \times 10^{-2}$$

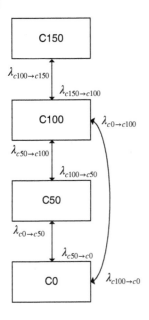

Figure 5.15 The equivalent state-transition diagram of the combined generation and transmission subsystem.

$$p_{c6} = p_{g100}p_{t100} = 2.664 \times 10^{-4}$$

$$p_{c7} = p_{g50}p_{t100} = 1.338 \times 10^{-6}$$

$$p_{c8} = p_{g0}p_{t100} = 2.242 \times 10^{-9}$$

$$p_{c9} = p_{g150}p_{t0} = 8.068 \times 10^{-5}$$

$$p_{c10} = p_{g100}p_{t0} = 1.216 \times 10^{-6}$$

$$p_{c11} = p_{g50}p_{t0} = 6.112 \times 10^{-9}$$

$$p_{c12} = p_{g0}p_{t0} = 1.024 \times 10^{-11}.$$

We can now simplify the above diagram by observing that the delivered capacity can only be 0, 50, 100 or 150 MW. The equivalent state-transition diagram for delivered capacity is shown in Figure 5.15.

The number of system states in generation subsystem reduces from 12 states to 4 states. Let the equivalent transition be λ; the equivalent transition rates can be found using (5.9) as follows. All units are in per year:

$$\lambda_{c150 \to c100} = \frac{p_{c1}(\lambda_{g150 \to g100} + \lambda_{t200 \to t100})}{p_{c1}} = 30.95$$

$$\lambda_{c100 \to c50} = \frac{p_{c2}\lambda_{g100 \to g50} + p_{c6}\lambda_{g100 \to g50}}{p_{c2} + p_{c5} + p_{c6}} = 333.36$$

$$\lambda_{c100 \to c0} = \frac{p_{c5}\lambda_{t100 \to t0} + p_{c6}\lambda_{t100 \to t0}}{p_{c2} + p_{c5} + p_{c6}} = 5.52$$

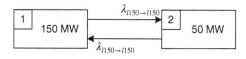

Figure 5.16 A state-transition diagram of a system load.

$$\lambda_{c50\to c0} = \frac{p_{c3}\lambda_{g50\to g0} + p_{c7}(\lambda_{g50\to g0} + \lambda_{t100\to t0})}{p_{c3} + p_{c7}} = 3.83$$

$$\lambda_{c0\to c50} = \frac{p_{c4}\lambda_{g0\to g50} + p_{c8}\lambda_{g0\to g50} + p_{c11}\lambda_{t0\to t100}}{p_{c4} + p_{c8} + p_{c9} + p_{c10} + p_{c11} + p_{c12}} = 3.50$$

$$\lambda_{c0\to c100} = \frac{p_{c9}\lambda_{t0\to t100} + p_{c10}\lambda_{t0\to t100}}{p_{c4} + p_{c8} + p_{c9} + p_{c10} + p_{c11} + p_{c12}} = 2186.50$$

$$\lambda_{c50\to c100} = \frac{p_{c3}\lambda_{g50\to g100} + p_{c7}(\lambda_{g50\to g100})}{p_{c3} + p_{c7}} = 1460$$

$$\lambda_{c100\to c150} = \frac{p_{c2}\lambda_{g100\to g150} + p_{c5}\lambda_{t100\to t200}}{p_{c2} + p_{c5} + p_{c6}} = 922.34.$$

Denote the probability $p_{c150}, p_{c100}, p_{50}, p_{c0}$ as the probability of the combined generation and transmission subsystem delivering the equivalent of 150, 100, 50 and 0 MW, respectively. The probability of being in each state in the equivalent diagram is found as follows:

$$p_{c150} = p_{c1} = 9.673 \times 10^{-1}$$

$$p_{c100} = p_{c2} + p_{c5} + p_{c6} = 3.252 \times 10^{-2}$$

$$p_{c50} = p_{c3} + p_{c7} = 7.462 \times 10^{-5}$$

$$p_{c0} = p_{c4} + p_{c8} + p_{c9} + p_{c10} + p_{c11} + p_{c12} = 8.202 \times 10^{-5}.$$

Load subsystem

The state-transition diagram of load is just a two-state Markov model, which is given in Figure 5.16.

The parameters of this diagram are given below:

$$\lambda_{l150\to l50} = \frac{1}{8} \text{ per hour} = 1095 \text{ per year}$$

$$\lambda_{l50\to l150} = \frac{1}{16} \text{ per hour} = 547.5 \text{ per year,}$$

where $\lambda_{l150\to l50}$ is the transition rate from the load of 150 to 50 MW, and $\lambda_{l50\to l150}$ is the transition rate from the load of 50 to 150 MW. The probabilities of the load equal to 150 and 50 MW, p_{l150}, p_{l50} are found as below:

$$p_{l150} = \frac{8}{8 + 16} = \frac{1}{3}$$

$$p_{l50} = \frac{16}{8 + 16} = \frac{2}{3}.$$

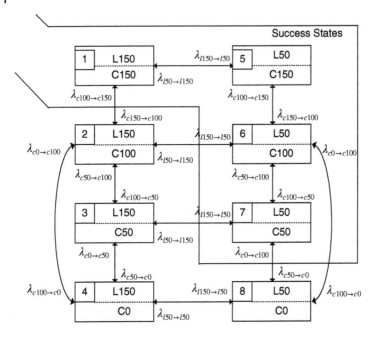

Figure 5.17 A system state-transition diagram.

Solution for the system

We can now find the solution to this problem by combining the state-transition diagram in Figure 5.15 with state-transition diagram of the system load shown in Figure 5.16. The system state-transition diagram of this problem is shown in Figure 5.17.

Using multiplication rule, we can find the probability of each system state as follows:

$$p_1 = p_{c150}p_{l150} = 3.224 \times 10^{-1}$$
$$p_2 = p_{c100}p_{l150} = 1.084 \times 10^{-2}$$
$$p_3 = p_{c50}p_{l150} = 2.487 \times 10^{-5}$$
$$p_4 = p_{c0}p_{l150} = 2.734 \times 10^{-5}$$
$$p_5 = p_{c150}p_{l50} = 6.449 \times 10^{-1}$$
$$p_6 = p_{c100}p_{l50} = 2.168 \times 10^{-2}$$
$$p_7 = p_{c50}p_{l50} = 4.975 \times 10^{-5}$$
$$p_8 = p_{c0}p_{l50} = 5.468 \times 10^{-5}.$$

It is now easy to find that the event of loss of load are the system states 2, 3 and 4. We can then find the loss of load probability as follows:

$$p_f = p_2 + p_3 + p_4 + p_8 = 1.095 \times 10^{-2}.$$

The frequency of loss of load in per year can be found from the diagram and is computed below:

$$f_f = p_1 \lambda_{c150 \to c100} + p_6(\lambda_{l50 \to l150} + \lambda_{c100 \to c0}) + p_7(\lambda_{l50 \to l150} + \lambda_{c50 \to c0}) = 22.$$

Lastly, we can find the mean down time in hours using (1.3) as follows:

$$T_D = \frac{p_f \times 24 \times 365}{f_f} = 4.36.$$

It can be seen from this example that, on average, loss of load events occur almost two times a month with the total average down time of 4.36 hours in one year.

5.3 Network Reduction Method

Network reduction method can be applied for systems whose interactions among components can be easily captured by a block diagram called *reliability block diagram*. The following assumptions need to be made before we can use the network reduction method to calculate reliability:

- Each component has only two states, up or down, and is represented by a block.
- Each component operates independently.
- The system can be represented by a simple reliability block diagram, and each block cannot be repeated.

However, it should be noted that in reality the method is used even when the blocks are repeated. In such a situation, when analytical expressions are developed, exact results can be obtained. In situations when only numerical calculations are made, the results may have some approximation. We first explain the concept of reliability block diagrams in the following section.

5.3.1 Reliability Block Diagrams

A reliability block diagram shows how failure of the components causes the system to fail. Each block represents one component whose condition can only be described as either working or failed. It can be considered as a switch whose on and off condition depends on whether the component is working or fails, respectively.

The system reliability block diagram is drawn by connecting these blocks in some fashion. When the failure of either one of the components causes the system to fail, we connect these components in *series structure*. When the failure of all components causes the system to fail, we connect these components in *parallel structure*. Examples of reliability block diagram are shown below.

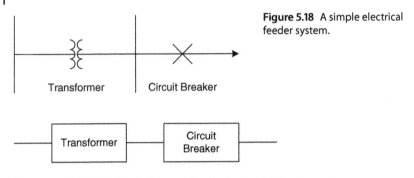

Figure 5.18 A simple electrical feeder system.

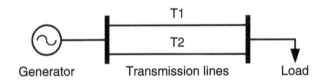

Figure 5.19 Reliability block diagram of a simple electrical feeder system.

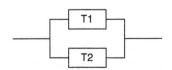

Figure 5.20 A system of two transmission lines in parallel.

Example 5.13   Consider a simple electrical feeder system that consists of one transformer and one circuit breaker, as shown in Figure 5.18.

Each component has two states, up or down, and is described by Figure 5.1. The system will fail when either one of the two components fails; thus this system has series structure. Let block A be a transformer and block B be a circuit breaker; we can draw the reliability block diagram as shown in Figure 5.19.

Example 5.14   Consider a system of two transmission lines in parallel as shown in Figure 5.20. The system fails only when both transmission lines fail. The reliability block diagram of this system is given by Figure 5.21. The system has a parallel structure.

We can see that the idea of a reliability block diagram is similar to a circuit diagram when failure of a component can be represented by an open circuit and when a working component is represented by a short circuit. In order for

Figure 5.21 Reliability block diagram of a system with two transmission lines.

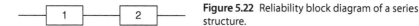

Figure 5.22 Reliability block diagram of a series structure.

the system to work, there should be at least one connected path. The system will fail when there is no connecting path.

5.3.2 Series Structure

Two components are said to be in series if the failure of either one causes system failure. An example of series structure is given in Example 5.13. It should be pointed out that it is not the physical layout but the effect on failure that defines the series system. Take two diodes in parallel for example: if one of the diodes breaks down in short circuit, the combination will fail. The failure mode of operation in this case is in series. A general block-diagram representation of a two-component series system is shown in Figure 5.22.

When the components are assumed to be independent, we can draw a state-transition diagram of the system, as in Figure 5.23, and identify the failure states of the system as state two, three and four.

We can now calculate the reliability indices of the series system as follows. The probability of success of this system is simply the multiplication of the probability of two components in the up state:

$$p_s = p_1 = \frac{\mu_1 \mu_2}{\lambda_1 \lambda_2 + \lambda_1 \mu_2 + \lambda_2 \mu_1 + \mu_1 \mu_2}. \tag{5.10}$$

We can easily find the probability of failure of this system by simply adding the probability of states two, three and four, or, equivalently,

$$p_f = p_2 + p_3 + p_4 = \frac{\lambda_1 \lambda_2 + \lambda_1 \mu_2 + \lambda_2 \mu_1}{\lambda_1 \lambda_2 + \lambda_1 \mu_2 + \lambda_2 \mu_1 + \mu_1 \mu_2}. \tag{5.11}$$

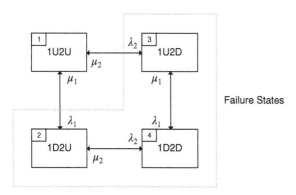

Figure 5.23 A state-transition diagram of two components in series.

The frequency of failure is the frequency from state one to states two, three and four, and is given by

$$f_f = p_1(\lambda_1 + \lambda_2), \tag{5.12}$$

where λ_i, μ_i are the failure rate and repair rate of component i.

We can also find the equivalent failure rate of the system, λ_{eq}, from success state one to failure states two, three and four by dividing the frequency of failure by the probability of success:

$$\lambda_{eq} = \frac{f_f}{p_s} = \lambda_1 + \lambda_2. \tag{5.13}$$

This means that we can now combine the two components in series and represent the system by one equivalent component whose failure rate is the summation of the two components' failure rates.

The equivalent repair rate of this system is found by dividing the frequency of success by the probability of failure. From the frequency-balance concept, we know that at the steady state the frequency of success is equal to frequency of failure. Thus, we can calculate the equivalent repair rate as follows:

$$\mu_{eq} = \frac{p_1(\lambda_1 + \lambda_2)}{1 - p_1} = \frac{\mu_1 \mu_2 (\lambda_1 + \lambda_2)}{\lambda_1 \mu_2 + \lambda_2 \mu_1 + \lambda_1 \lambda_2}. \tag{5.14}$$

Mean down time of this equivalent component can be found from the reciprocal of the repair rate. Let the down time of component i be r_i; then we have

$$T_d = \frac{1}{\mu_{eq}} = \frac{\lambda_1 r_1 + \lambda_2 r_2 + \lambda_1 \lambda_2 r_1 r_2}{\lambda_1 + \lambda_2}. \tag{5.15}$$

We can use (5.13), (5.14) and (5.15) to combine two independent components connected in series structure into one equivalent component. The above analysis is for independent failures. We discuss the special case of dependent failure in series structure in Example 5.15.

Example 5.15   Consider a system of two components as shown in Example 5.10. When the failures of both components in the series structure are dependent, once one component fails, the system is de-energized; thus the failure rate of the second component changes, and here we assume it drops to zero and further component failures will not thus occur. The state-transition diagram of dependent failure is shown in Figure 5.6. Components are no longer independent, so the state probabilities cannot be found by simple multiplication but from state space approach as shown in Example 5.10. Let us find the equivalent repair rate, equivalent failure rate and mean down time of this system.

The probability of success is the probability of state one, which is given again in (5.16):

$$p_s = p_1 = \frac{\mu_1 \mu_2}{\lambda_1 \mu_2 + \lambda_2 \mu_1 + \mu_1 \mu_2},\tag{5.16}$$

and the probability of failure is given below:

$$p_f = \frac{\lambda_1 \mu_2 + \lambda_2 \mu_1}{\lambda_1 \mu_2 + \lambda_2 \mu_1 + \mu_1 \mu_2}.\tag{5.17}$$

The frequency of failure is found from (5.18):

$$f_f = p_1(\lambda_1 + \lambda_2) = \frac{\lambda_1 + \lambda_2}{1 + \lambda_1 r_1 + \lambda_2 r_2}.\tag{5.18}$$

We can find the equivalent failure rate of the system as follows:

$$\lambda_{eq} = \frac{f_f}{p_s} = \lambda_1 + \lambda_2.\tag{5.19}$$

The equivalent repair rate of this system is found by dividing the frequency of success by the probability of failure:

$$\mu_{eq} = \frac{\mu_1 \mu_2 (\lambda_1 + \lambda_2)}{\lambda_1 \mu_1 + \lambda_2 \mu_1}.\tag{5.20}$$

Mean down time of this equivalent component can be found from the reciprocal of the repair rate:

$$T_d = \frac{1}{\mu_{eq}} = \frac{\lambda_1 r_1 + \lambda_2 r_2}{\lambda_1 + \lambda_2}.\tag{5.21}$$

It can be seen that the equivalent failure rates of both cases are the same. The differences between dependent and independent failure of the two components in series are in the state probability and equivalent repair rate/mean down time. The equations describing probability of failure of the two cases are different by the factor $\lambda_1 \lambda_2$. Similarly, the equations describing equivalent repair rate/mean down time of both cases are different by the factor $\lambda_1 \lambda_2 r_1 r_2$.

Generally, both $\lambda_1 \lambda_2$ and $\lambda_1 \lambda_2 r_1 r_2$ are rather small and can be ignored in comparison to the other terms. Although the results of both independent and dependent failure of the two components in series are different conceptually, numerically they are almost the same for most practical situations.

In the following section, we analyze the parallel structure and describe the equations for combining two components in parallel.

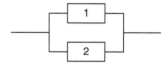

Figure 5.24 Reliability block diagram of a parallel structure.

5.3.3 Parallel Structure

Two components are said to be in parallel if both must fail to cause system failure. It should be, however, remembered that the components do not have to be physically in parallel. An example of two parallel circuits of transmission lines is shown in Example 5.14. A general block diagram representation of a two-component parallel system is shown in Figure 5.24.

When the components are assumed to be independent, we can draw a state-transition diagram of the system as shown in Figure 5.25 and identify state four as the failure state of the system. Note that the state-transition diagram of the parallel system is exactly the same as that of the series system. The difference is that the parallel structure has only one failure state.

In this case, we can find the failure probability simply by multiplying the failure probability of components one and two:

$$p_f = p_4 = \frac{\lambda_1 \lambda_2}{\lambda_1 \lambda_2 + \lambda_1 \mu_2 + \lambda_2 \mu_1 + \mu_1 \mu_2}. \tag{5.22}$$

The probability of success of this system is found from adding the probability of states two, three and four, or, equivalently, the complement of the failure probability:

$$p_s = p_1 + p_2 + p_3 = \frac{\lambda_1 \mu_2 + \lambda_2 \mu_1 + \mu_1 \mu_2}{\lambda_1 \lambda_2 + \lambda_1 \mu_2 + \lambda_2 \mu_1 + \mu_1 \mu_2}. \tag{5.23}$$

The frequency of failure can be found as the expected transition rate from success state (states one, two and three) to failure state (state four).

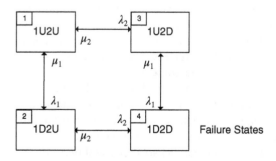

Figure 5.25 A state-transition diagram of two components in parallel.

Alternatively, frequency of failure is equal to frequency of success at steady state, which is the expected transition rate from failure state (state four) to success states (states one, two and three), which is easier to calculate:

$$f_f = f_s = p_4(\mu_1 + \mu_2). \tag{5.24}$$

We can also find equivalent repair rate by dividing the frequency of success by the probability of failure:

$$\mu_{eq} = \frac{f_s}{p_f} = \mu_1 + \mu_2. \tag{5.25}$$

Mean down time of this equivalent component can be found from the reciprocal of the repair rate. Let the down time of component i be r_i; then we have

$$T_d = \frac{1}{\mu_{eq}} = \frac{r_1 r_2}{r_1 + r_2}. \tag{5.26}$$

The equivalent failure rate of this system can also be found by dividing the frequency of failure by the probability of success as follows:

$$\lambda_{eq} = \frac{p_4(\mu_1 + \mu_2)}{1 - p_4} = \frac{\lambda_1 \lambda_2(\mu_1 + \mu_2)}{\lambda_1 \mu_2 + \lambda_2 \mu_1 + \mu_1 \mu_2} = \frac{\lambda_1 \lambda_2(r_1 + r_2)}{1 + \lambda_1 r_1 + \lambda_2 r_2}. \tag{5.27}$$

We can use (5.25), (5.26) and (5.27) to combine two independent components connected in parallel structure into one equivalent component.

Some structures are not either series or parallel; an example of such structure is *bridge structure* and is shown in Example 5.16.

Example 5.16   Consider a system with two distribution feeders where both feeders are connected via a circuit breaker that is normally open, as shown in Figure 5.26. During the failure of either one of the distribution feeder lines, this circuit breaker can reconnect the failed feeder to the working feeder to supply electricity to the load of the failed feeder. We can represent this system using the reliability block diagram, as shown in Figure 5.27, where components 1, 2, 3, 4 and 5 represents transformer 1, transformer 2, circuit breaker 1, circuit breaker 2 and circuit breaker 3, respectively.

Although methods like star-delta transformation have been developed to convert the more complicated structures into series-parallel structures, by far network reduction method is used mostly to reduce such systems in series and/or parallel structures. The objective of the network reduction approach is to combine two components in parallel or series into an equivalent component.

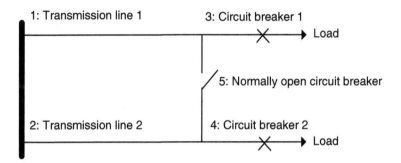

Figure 5.26 A system with two distribution feeders.

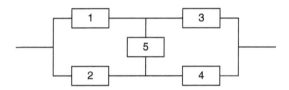

Figure 5.27 Reliability block diagram of a bridge-structure system.

5.3.4 Steps of Network Reduction Approach

The previous sections describe the equivalent failure rate, repair rate and mean down time of the two-component series and parallel systems. In this section, we describe the process of network reduction method as a step-by-step procedure below:

Step 1 Identify series-parallel structures in reliability block diagram.

Step 2 Replace all series blocks by equivalent components and replace all parallel blocks by equivalent components.

Step 3 Repeat the above steps until the whole network reduces to an equivalent component.

The system indices are found from the reliability indices of the equivalent component. We illustrate this procedure in the following example.

Example 5.17 A system consists of one component in series with two components in parallel as shown in Figure 5.28. The failure rate of each component is 0.05 failure per year and the mean repair time is 10 hours. Using

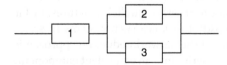

Figure 5.28 Reliability block diagram of the system in Example 5.17.

the network reduction method, calculate the failure rate and the mean down time of this system.

First we can combine the components two and three that are in parallel. In this case, $\lambda_1 = \lambda_2 = \lambda_3 = 0.05$ failures per year or, equivalently, 5.708×10^{-6} failures per hour, $r_1 = r_2 = r_3 = 10$ hours. Using (5.27), the equivalent failure rate of this combination is

$$\lambda_{23} = \frac{5.708 \times 10^{-6} \times 5.708 \times 10^{-6} \times (10 + 10)}{1 + 5.708 \times 10^{-6} \times 10 + 5.708 \times 10^{-6}}$$

$$= 6.515 \times 10^{-10} \text{ failures per hour}$$

$$= 5.707 \times 10^{-6} \text{ failures per year,}$$

and the equivalent mean down time is found using (5.26):

$$d_{23} = \frac{10 \times 10}{20} = 5 \text{ hours.}$$

Then this equivalent can be combined with component one in series; we can find the equivalent failure rate and mean down time of the system using (5.13) and (5.15):

$$\lambda_{123} = \lambda_1 + \lambda_{23}$$

$$= 0.05 + 5.707 \times 10^{-6}$$

$$\approx 0.05 \text{ failures per year}$$

$$d_{123} = \frac{5.707 \times 10^{-6} \times 5 + 0.05 \times 10 + 5.707 \times 10^{-6} \times 0.05 \times 10 \times 5}{0.05 + 5.707 \times 10^{-6}}$$

$$= 9.999 \text{ hours}$$

$$\approx 10 \text{ hours.}$$

The network reduction approach can be conveniently used for systems in series-parallel structures. Some systems may not be categorized as either series or parallel, including, for example, the bridge structure shown in Figure 5.27. For this particular system, when component five fails, components one and two are in series structure, as are components three and four. When component five is working, components one and three are in parallel structure as are components two and four. This means that we can analyze the system structure easily when we know the status of component five. We call this method *conditional probability method*, which is explained in the next section.

5.4 Conditional Probability Method

For some complex systems that cannot be classified as either series or parallel structures, we can identify key components whose status can reduce a reliability block diagram to a combination of series and parallel structures. The fundamental concept of this method is *conditional probability* from Chapter 2 and *conditional frequency* from Chapter 4, which are used to calculate reliability indices of this network. We review both concepts again in this section.

In this application, we use the conditional probability concept to calculate probability of a failure event, Y. We first identify a key component in the system. Let component k be a key component, the knowledge of whose status can help to simplify the overall system structure. We denote an event K as the event that the component k is working and an event $\bar{K}$ as the event that the key component k fails. This means that we can partition the state space S into two disjoint sets, K and $\bar{K}$, $K \cap \bar{K} = \emptyset$, $K \cup \bar{K} = S$. We can find probability of a failure event Y as follows:

$$P(Y) = P(Y|K) \times P(K) + P(Y|\bar{K}) \times P(\bar{K}). \tag{5.28}$$

Once the status of the key component k is known, the structure of the system can be simplified as a combination of series and parallel structures. We can then use network reduction method to find the failure probability of the corresponding structure.

Similarly, we can use the conditional frequency concept to find frequency of failure as follows:

$$Fr_{\to Y} = Fr_{(S\backslash Y)|K \to Y|K} + Fr_{(S\backslash Y)|\bar{K} \to Y|\bar{K}} + Fr_{(S\backslash Y)|K \to Y|\bar{K}}, \tag{5.29}$$

where $Fr_{(S\backslash Y)|K \to Y|K}$ is the frequency of failure when component k is working. $Fr_{(S\backslash Y)|\bar{K} \to Y|\bar{K}}$ is the frequency of failure when component k is failed. $Fr_{(S\backslash Y)|K \to Y|\bar{K}}$ is the frequency of failure as a result of the change in component k's status, which is derived in Section 4.7 and is given again below:

$$Fr_{(S\backslash Y)|K \to Y|\bar{K}} = (P\{(S \backslash Y)|K\} - P\{(S \backslash Y)|\bar{K}\}) \times p_k \lambda_k, \tag{5.30}$$

where p_k is the probability of success of component k, and λ_k is the failure rate of component k.

At steady state, the frequency of failure is the same as frequency of success. We can also calculate frequency of failure from frequency of success as follows:

$$Fr_{Y \to} = Fr_{Y|K \to (S\backslash Y)|K} + Fr_{Y|\bar{K} \to (S\backslash Y)|\bar{K}} + Fr_{Y|\bar{K} \to (S\backslash Y)|K}, \tag{5.31}$$

where $Fr_{Y|\bar{K} \to (S\backslash Y)|K}$ is the frequency of success as a result of the change in component k's status, which is derived in Section 4.7 and is given again below:

$$Fr_{Y|\bar{K} \to (S\backslash Y)|K} = (P\{Y|\bar{K}\} - P\{Y|K\}) \times p_{\bar{k}} \mu_k, \tag{5.32}$$

where $p_{\bar{k}}$ is the probability of failure of component k, and μ_k is the repair rate of component k.

The conditional probability and frequency concept is depicted by Figure 5.29.

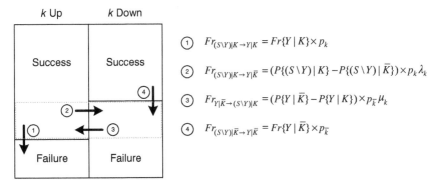

k Up k Down

① $Fr_{(S\setminus Y)|K \to Y|K} = Fr\{Y \mid K\} \times p_k$

② $Fr_{(S\setminus Y)|K \to Y|\bar{K}} = (P\{(S\setminus Y) \mid K\} - P\{(S\setminus Y) \mid \bar{K}\}) \times p_k \lambda_k$

③ $Fr_{Y|\bar{K} \to (S\setminus Y)|K} = (P\{Y \mid \bar{K}\} - P\{Y \mid K\}) \times p_{\bar{k}} \mu_k$

④ $Fr_{(S\setminus Y)|\bar{K} \to Y|\bar{K}} = Fr\{Y \mid \bar{K}\} \times p_{\bar{k}}$

Figure 5.29 Conditional probability and frequency concept for a complex system with a key component k.

We illustrate the concept of the conditional probabiltiy method in the following example.

Example 5.18 For a bridge structure shown in Figure 5.27, each component has a failure rate of λ per year and a repair rate of μ per year. Find the probability of failure, frequency of failure, equivalent repair rate and equivalent failure rate of the whole system using network reduction method together with conditional probability method.

From the failure and repair rate, the failure probability of each component, p_d is simply:

$$p_d = \frac{\lambda}{\lambda + \mu}.$$

When the component five is down, the system is simplified to Figure 5.30. Let us denote this system system A.

We can see that now the components one and three are in series, as are components two and four. We can find the failure probability of system A as follows:

$$p_f^A = (1 - (1 - p_d)^2)^2 = \frac{\lambda^2(\lambda + 2\mu)^2}{(\lambda + \mu)^4}.$$

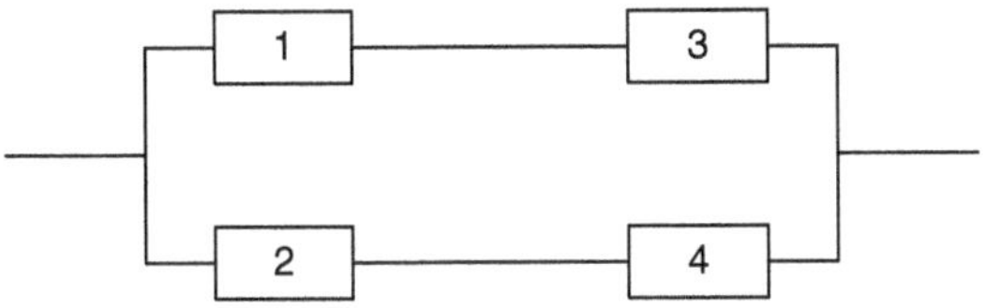

Figure 5.30 Bridge network in Example 5.18 when component five is down.

We can also find the frequency of failure of system A using the network reduction method. Since components one and three are in series, we can find their equivalent repair rate using (5.14):

$$\mu_{1\&3} = \frac{2\mu^2}{\lambda + 2\mu}.$$

Components two and four are in series, and their equivalent repair rate of is $\mu_{2\&4} = \mu_{1\&3}$. Now the components one and three are in parallel to components two and four we should find equivalent repair rate of system A as follows:

$$\mu_{eq}^A = \mu_{1\&3} + \mu_{2\&4} = \frac{4\mu^2}{\lambda + 2\mu}.$$

The frequency of failure of system A is found from

$$f_f^A = p_f^A \times \mu_{eq}^A = \frac{4\lambda^2\mu^2(\lambda + 2\mu)}{(\lambda + \mu)^4}.$$

When the component five is up, the system is simplified to Figure 5.31. Let us denote this system system B.

We can see that now the components one and two are in parallel, as are components three and four. We can find the failure probability of system B as follows:

$$p_f^B = 1 - (1 - p_d^2)^2 = \frac{\lambda^2(\lambda^2 + 4\lambda\mu + 2\mu^2)}{(\lambda + \mu)^4}.$$

We can also find the frequency of failure of system B using the network reduction method. Since components one and two are in parallel, we can find their equivalent failure rate using (5.27):

$$\lambda_{1\&2} = \frac{2\lambda^2}{2\lambda + \mu}.$$

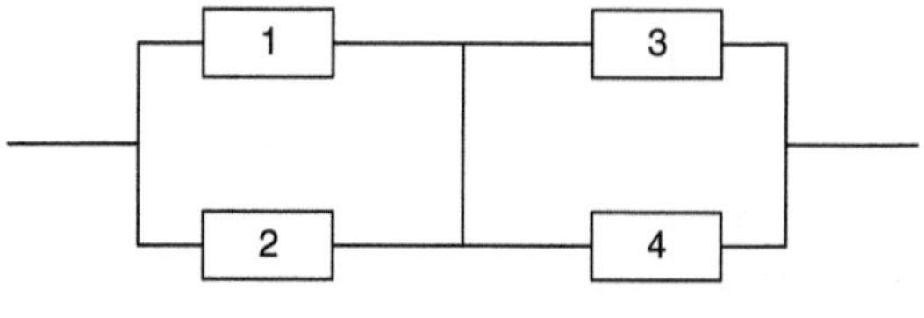

Figure 5.31 Bridge network in Example 5.18 when component five is up.

Components three and four are in parallel, and their equivalent failure rate is $\lambda_{3\&4} = \lambda_{1\&2}$. Now components one and two are in parallel to components three and four, we can find equivalent failure rate of system B as follows:

$$\lambda_{eq}^B = \lambda_{1\&2} + \lambda_{3\&4} = \frac{4\lambda^2}{2\lambda + \mu}.$$

The frequency of failure of system B is found from:

$$f_f^B = f_s^B = p_s^B \times \lambda_{eq}^B = \frac{4\lambda^2 \mu^2 (2\lambda + \mu)}{(\lambda + \mu)^4}.$$

Using (5.28), we can find the failure probability as follows:

$$p_f = P\{\text{failure}|5\text{ down}\} \times P\{5\text{ down}\} + P\{\text{failure}|5\text{ up}\} \times P\{5\text{ up}\}$$

$$= p_f^A \times \frac{\lambda}{\lambda + \mu} + p_f^B \times \frac{\mu}{\lambda + \mu}$$

$$= \frac{\lambda^2 (\lambda^3 + 5\lambda^2 \mu + 8\lambda\mu^2 + 2\mu^3)}{(\lambda + \mu)^5}.$$

Using (5.29) and (5.30), we can find the failure frequency as follows:

$$f_f = Fr_{(S\backslash Y)|K \to Y|K} + Fr_{(S\backslash Y)|\bar{K} \to Y|\bar{K}} + Fr_{(S\backslash Y)|K \to Y|\bar{K}}$$

$$= f_f^A \times \frac{\lambda}{\lambda + \mu} + f_f^B \times \frac{\mu}{\lambda + \mu} + ((1 - p_f^B) - (1 - p_f^A)) \times (1 - p_d)\lambda$$

$$= \frac{2\lambda^2 \mu^2 (2\lambda^2 + 9\lambda\mu + 2\mu^2)}{(\lambda + \mu)^5}.$$

Similarly, we can use (5.31) and (5.32) to find frequency of success, which is the same as frequency of failure at steady state, as shown below:

$$f_s = Fr_{Y|K \to (S\backslash Y)|K} + Fr_{Y|\bar{K} \to (S\backslash Y)|\bar{K}} + Fr_{Y|\bar{K} \to (S\backslash Y)|K}$$

$$= f_s^A \times \frac{\lambda}{\lambda + \mu} + f_s^B \times \frac{\mu}{\lambda + \mu} + (p_f^A - p_f^B)) \times p_d \mu$$

$$= \frac{2\lambda^2 \mu^2 (2\lambda^2 + 9\lambda\mu + 2\mu^2)}{(\lambda + \mu)^5}.$$

It is shown in Example 5.18 that, with the help of conditional probability and frequency, we can use the network reduction method for a more complicated system whose configurations are neither in series nor parallel structure. In the following section we discuss an alternative method to calculate reliability indices of a complicated structure, called *cut-set and tie-set method*.

5.5 Cut-Set and Tie-Set Methods

The main idea of cut-set and tie-set methods is to represent a reliability block diagram of a complicated structure by an equivalent one in a combination of series and parallel structures. The resulting equivalent block diagram may show repeated blocks of the same device, which violates the assumption of the network reduction method. In the following sections, we describe the concept of each method in detail and derive reliability indices in terms of probability of failure and frequency of failure using each method.

5.5.1 Cut-Set Method

We first explain the definition of a cut set and a minimal (min) cut set.

Definition 5.1 A cut set *is set of components whose failure will cause system failure.*

Definition 5.2 A minimal cut set, *denoted by C, is a cut set that has no proper subset as a cut set except itself.*

In other words, the minimal cut set contains a combination of components whose failure causes system to fail. The term component is used here in a general sense; it could be interpreted as a physical component or existence of a condition. By taking this broader interpretation of component, the method becomes a very powerful tool for reliability analysis. The order of a cut set indicates the number of components in the cut set.

Example 5.19 Find the min cut sets for the system in Figure 5.27.
For this system, the minimum cut sets are $C_1 = \{1, 2\}, C_2 = \{4, 5\}, C_3 = \{1, 3, 5\}$ and $C_4 = \{2, 3, 4\}$.
The main difference between the cut set and minimum cut set should be noted. For example, in this system $\{1, 2, 4\}$ is a cut set, but since $\{1,2\}$, a subset of this set, is also a cut set, it is not a minimum cut set. In this example C_1 and C_2 are second-order cut sets and C_3 and C_4 are third-order cuts sets.

After the min cut sets have been identified, there are two basic methods for computing the reliability indices: direct method and block diagram representation method. Both of these methods are explained next.

Reliability Indices Calculation Using Direct Method
Recall that each min cut set represents an event whose failure of all components in the set causes a system to fail. This means that failure of the system is caused by the failure of all components in any min cut set.

Let us denote the event of failure of all components in C_i $\bar{C}_i$, and Y be the failure event of the system. We can write

$$Y = \bar{C}_1 \cup \bar{C}_2 \cup \bar{C}_3 \ldots \cup \bar{C}_n.$$

The probability of failure can be simply computed using the expression for the union of events from Chapter 2 as follows:

$$p_f = P(\bar{C}_1 \cup \bar{C}_2 \cup \bar{C}_3 \ldots \cup \bar{C}_n) \tag{5.33}$$

$$= \sum_i P(\bar{C}_i) - \sum_{i<j} P(\bar{C}_i \cap \bar{C}_j) \tag{5.34}$$

$$+ \sum_{i<j<k} P(\bar{C}_i \cap \bar{C}_j \cap \bar{C}_k) - \ldots \tag{5.35}$$

$$+(-1)^{n-1} P(\bar{C}_1 \cap \bar{C}_2 \cap \ldots \cap \bar{C}_n). \tag{5.36}$$

We have a similar expression for frequency calculation from Chapter 4. Let us first recall from (4.16) that $\forall X, Y \subset S, X \cap Y = \emptyset$, and when $Y = \bar{C}_1 \cup \bar{C}_2 \cup \bar{C}_3 \ldots \cup \bar{C}_n$, then

$$Fr_{X \to Y} = \sum_i Fr_{X \to \bar{C}_i} - \sum_{i<j} Fr_{X \to \bar{C}_i \cap \bar{C}_j}$$

$$+ \sum_{i<j<k} Fr_{X \to \bar{C}_i \cap \bar{C}_j \cap \bar{C}_k} - \ldots$$

$$+(-1)^{n-1} Fr_{X \to \bar{C}_1 \cap \bar{C}_2 \cap \ldots \cap \bar{C}_n}.$$

From the above expression, we can let X be the set of success states; this means that we can then calculate the frequency of entering failure states from success states or system frequency of success. Alternatively, we can compute the frequency of success from the probability of failure in (5.36) using the conversion rule given in Section 4.9, which is given again below:

$$Fr_{Y \to} = \sum_{i \in Y} p_i \left(\sum_{k \in \bar{K}_i} \mu_k - \sum_{k \in K_i} \frac{p_{\bar{k}} \mu_k}{p_k} \right),$$

where p_i is probability of state i, $\bar{K}_i$ is the set of components that is down in system state i, K_i is the set of components that is up in system state i, p_k is the probability of component k in up state, $p_{\bar{k}}$ is the probability of component k in down state, and $i\mu_k$ is a repair rate of component k.

Since the components in the min cut set $\bar{C}_i$ are all in failure states, we can simplify the above expression as follows:

$$Fr_{Y \to} = \sum_{i \in Y} p_i \left(\sum_{k \in \bar{K}_i} \mu_k \right).$$

Using the above conversion rules, the probability expressions given in (5.36) can be easily converted into frequency of success. Basically, each term in the probability expression is multiplied by the summation of repair rates of components that are failed in that term. Thus, frequency of success is given by (5.37):

$$f_s = \sum_{i=1}^{n} P(\bar{C}_i)\bar{\mu}_i - \sum_{i<j}^{n} P(\bar{C}_i \cap \bar{C}_j)\bar{\mu}_{i+j} \tag{5.37}$$

$$+ \sum_{i<j<k}^{n} P(\bar{C}_i \cap \bar{C}_j \cap \bar{C}_k)\bar{\mu}_{i+j+k} - \cdots \tag{5.38}$$

$$+(-1)^n P(\bar{C}_1 \cap \bar{C}_2 \cap \bar{C}_3 \cap \bar{C}_4 \cap \ldots \cap \bar{C}_n)\bar{\mu}_{i+j+\ldots+n}, \tag{5.39}$$

where $\bar{\mu}_i$ is summation of μ_i of all components $\in C_i$, and

$$\bar{\mu}_{i+j+\ldots+n} = \sum_{m \in C_i \cup C_j \cup C_k \cup \ldots \cup C_n} \mu_m \tag{5.40}$$

At steady state, the frequency of success and frequency of failure are the same, so we can use (5.37) to find system frequency of failure. Once the probability and frequency have been calculated, the mean duration can be computed. The various terms in these expressions can be explained by using the following min cut sets from a bridge system in Example 5.19.

Example 5.20   Let us consider two min cuts $C_1 = \{1, 2\}$, $C_3 = \{1, 3, 5\}$, using an upper bar to indicate failure of the component in each cut set.

$$\bar{C}_1 = \bar{1}, \bar{2}$$
$$\bar{C}_3 = \bar{1}, \bar{3}, \bar{5}$$
$$\bar{C}_1 \cap \bar{C}_3 = \bar{1}, \bar{2}, \bar{3}, \bar{5}$$

The equivalent repair rates of each cut set can be found from (5.48).

$$\bar{\mu}_1 = \mu_1 + \mu_2$$
$$\bar{\mu}_3 = \mu_1 + \mu_3 + \mu_5$$
$$\bar{\mu}_{1+3} = \mu_1 + \mu_2 + \mu_3 + \mu_5$$

The probability and frequency of failure from each cut set are found below:

$$P(\bar{C}_1) = p_{1d}p_{2d}$$
$$P(\bar{C}_3) = p_{1d}p_{3d}p_{5d}$$
$$P(\bar{C}_1 \cap \bar{C}_3) = p_{1d}p_{2d}p_{3d}p_{5d}$$
$$Fr_{\to \bar{C}_1} = P(\bar{C}_1) \times \bar{\mu}_1 = p_{1d}p_{2d}(\mu_1 + \mu_2)$$
$$Fr_{\to \bar{C}_3} = P(\bar{C}_3) \times \bar{\mu}_3 = p_{1d}p_{3d}p_{5d}(\mu_1 + \mu_3 + \mu_5)$$
$$Fr_{\to (\bar{C}_1 \cap \bar{C}_3)} = P(\bar{C}_1 \cap \bar{C}_3) \times \bar{\mu}_{1+3} = p_{1d}p_{2d}p_{3d}p_{5d}(\mu_1 + \mu_2 + \mu_3 + \mu_5).$$

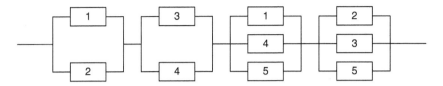

Figure 5.32 Equivalent reliability block diagram of a bridge-structure system.

Reliability Indices Calculation Using Block Diagrams

Since all components in a min cut set need to fail to cause system failure, the components in a min cut set can be regarded as connected in parallel structure. As any min cut set can cause system failure, the min cut sets are connected in series structure. We can now rearrange the complicating structures as a combination of series and parallel structures using the cut set method. This can be illustrated in the following example.

Example 5.21 Consider a bridge system shown in Figure 5.27. The min cuts of this system are found in Example 5.19, which are {1,2}, {4,5}, {1,3,5} and {2,3,4}. We can draw an equivalent reliability block diagram from these min cut sets as seen in Figure 5.32.

We can use the network reduction method to successively reduce the parallel and series combinations into an equivalent component representing the system to find reliability indices. It should, however, be noted that results obtained by this method would be correct only if there is no repeated block in the diagram. Consider the equivalent block diagram shown in Example 5.21; we can see that blocks are repeated. For example, block 1 is repeated in cut sets 1 and 3, and thus the results will not be exact. We illustrate this point in the following example.

Example 5.22 Let us calculate the probability of failure of an equivalent block diagram shown in Figure 5.32; we have:

$$p_f = 1 - (1 - p_{1d}p_{2d})(1 - p_{4d}p_{5d})(1 - p_{1d}p_{3d}p_{5d})(1 - p_{2d}p_{3d}p_{4d})$$
$$= p_{1d}p_{2d} + p_{4d}p_{5d} - p_{1d}p_{2d}p_{4d}p_{5d} + p_{1d}p_{3d}p_{5d}$$
$$- p_{1d}p_{3d}p_{4d}p_{5d}^2 - p_{1d}^2 p_{2d}p_{3d}p_{5d} + p_{1d}^2 p_{2d}p_{3d}p_{4d}p_{5d}^2$$
$$+ p_{2d}p_{3d}p_{4d} - p_{2d}p_{3d}p_{4d}^2 p_{5d} - p_{1d}p_{2d}^2 p_{3d}p_{4d}$$
$$+ p_{1d}p_{2d}^2 p_{3d}p_{4d}^2 p_{5d} - p_{1d}p_{2d}p_{3d}^2 p_{4d}p_{5d} + p_{1d}p_{2d}p_{3d}^2 p_{4d}^2 p_{5d}^2$$
$$+ p_{1d}^2 p_{2d}^2 p_{3d}^2 p_{4d}p_{5d} - p_{1d}^2 p_{2d}^2 p_{3d}^2 p_{4d}^2 p_{5d}^2.$$

We can see from this expression that some terms have an exponent greater than one. For example, the fifth term has the square of the p_{5d}. This is obviously meaningless, as it indicates as it is the joint probability of component 5 failing with component 5. This expression can be easily corrected, considering the fact that the joint probability of the component 5 failing with component 5 is only p_{5d}. We can now correct the probability of failure expression by setting all exponents to one and the resulting expression will be exact. For this case it is:

$$p_f = p_{1d}p_{2d} + p_{4d}p_{5d} + p_{1d}p_{3d}p_{5d} + p_{2d}p_{3d}p_{4d}$$
$$- (p_{1d}p_{2d}p_{4d}p_{5d} + p_{1d}p_{3d}p_{4d}p_{5d} + p_{1d}p_{2d}p_{3d}p_{5d} + p_{2d}p_{3d}p_{4d}p_{5d}$$
$$+ p_{1d}p_{2d}p_{3d}p_{4d}) + 2p_{1d}p_{2d}p_{3d}p_{4d}p_{5d}.$$

These are the same terms we obtain by expanding the expression for p_f, discussed under the direct method.

We can now find the frequency of failure from Chapter 4; we have:

$$f_f = f_{\bar{C}_1 \cup \bar{C}_2 \cup \bar{C}_3 \ldots \cup \bar{C}_n} \tag{5.41}$$

$$= \sum_i f_{\bar{C}_i} - \sum_{i<j} f_{\bar{C}_i \cap \bar{C}_j} \tag{5.42}$$

$$+ \sum_{i<j<k} f_{\bar{C}_i \cap \bar{C}_j \cap \bar{C}_k} - \cdots \tag{5.43}$$

$$+(-1)^{n-1} f_{\bar{C}_1 \cap \bar{C}_2 \cap \ldots \cap \bar{C}_n}, \tag{5.44}$$

where $f_{\bar{C}}$ is the frequency of failure as a result of cut set $\bar{C}$.

From the above expression, we need to find the frequency of failure of cut sets $\bar{C}_i$, $\bar{C}_i \cap \bar{C}_j$, $\bar{C}_i \cap \bar{C}_j \cap \bar{C}_k$, and so on. Since all components in the cut set are connected in parallel, it is easier to find the frequency of success by multiplying the probability of failure of each cut set with its equivalent repair rate. From the network reduction method, we know that the equivalent repair rate is simply found from summation of the repair rate of all components. We can now easily calculate the frequency of failure as follows:

$$f_f = \sum_{i=1}^n P(\bar{C}_i)\bar{\mu}_i - \sum_{i<j}^n P(\bar{C}_i \cap \bar{C}_j)\bar{\mu}_{i+j} \tag{5.45}$$

$$+ \sum_{i<j<k}^n P(\bar{C}_i \cap \bar{C}_j \cap \bar{C}_k)\bar{\mu}_{i+j+k} - \cdots \tag{5.46}$$

$$+(-1)^n P(\bar{C}_1 \cap \bar{C}_2 \cap \bar{C}_3 \cap \bar{C}_4 \cap \ldots \cap \bar{C}_n)\bar{\mu}_{i+j+\ldots+n}, \tag{5.47}$$

where $\bar{\mu}_i$ is the equivalent repair rate found from the summation of μ_i of all components $\in C_i$ and

$$\bar{\mu}_{i+j+\ldots+n} = \sum_{m \in C_i \cup C_j \cup C_k \cup \ldots \cup C_n} \mu_m. \tag{5.48}$$

We arrive at the same formula as found from direct method using conversion rules. The following example shows how to calculate the frequency of failure from probability of failure.

Example 5.23   Let us calculate the frequency of failure of an equivalent block diagram shown in Figure 5.32. We have:

$$
\begin{aligned}
f_f &= p_{1d}p_{2d}(\mu_1 + \mu_2) + p_{4d}p_{5d}(\mu_4 + \mu_5) + p_{1d}p_{3d}p_{5d}(\mu_1 + \mu_3 + \mu_5) \\
&+ p_{2d}p_{3d}p_{4d}(\mu_2 + \mu_3 + \mu_4) - (p_{1d}p_{2d}p_{4d}p_{5d}(\mu_1 + \mu_2 + \mu_4 + \mu_5) \\
&+ p_{1d}p_{3d}p_{4d}p_{5d}(\mu_1 + \mu_3 + \mu_4 + \mu_5) + p_{1d}p_{2d}p_{3d}p_{5d}(\mu_1 + \mu_2 + \mu_3 + \mu_5) \\
&+ p_{2d}p_{3d}p_{4d}p_{5d}(\mu_2 + \mu_3 + \mu_4 + \mu_5) + p_{1d}p_{2d}p_{3d}p_{4d}(\mu_1 + \mu_2 + \mu_3 + \mu_4)) \\
&+ 2p_{1d}p_{2d}p_{3d}p_{4d}p_{5d}(\mu_1 + \mu_2 + \mu_3 + \mu_4 + \mu_5).
\end{aligned}
$$

As has been demonstrated in the two above examples, if symbolic expressions can be written for the probability of failure, exact values for probability and frequency of system failure can be determined using the method of block representation of the min cut sets. However, this is possible in only small systems. For relatively larger systems, *method of sequential reduction of series and parallel structures need to be used, and in this process only numerical values at each step are retained. So the results obtained can be only approximate.*

Bounds of Reliability Indices

For large systems, the computational burden could become impractical due to the large number of terms involved. If there are n min cut sets, the number of terms is 2^n. The computational burden could be reduced in two ways. First, the min cut sets of only up to a certain order can be used, as the probability of failure of higher-order cuts sets becomes very small. Second, the bounds can be used instead of the final result.

The first upper and lower bounds on the probability and frequency of failure are given below:

$$p_U = \text{First upper bound to } p_f$$

$$= \sum_i P(\bar{C}_i)$$

$$p_L = \text{First lower bound to } p_f$$

$$= p_U - \sum_{i<j} P(\bar{C}_i \cap \bar{C}_j)$$

$$f_U = \text{First upper bound to } f_f$$

$$= \sum_i P(\bar{C}_i)\mu_i$$

$$f_L = \text{First lower bound to } f_f$$

$$= f_U - \sum_{i<j} P(\bar{C}_i \cap \bar{C}_j)\mu_{i+j}.$$

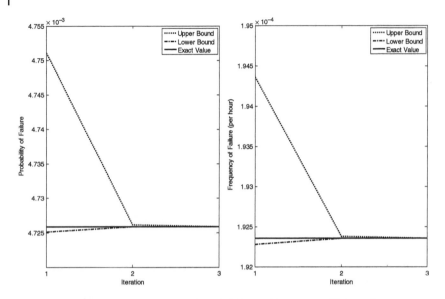

Figure 5.33 Lower and upper bounds of probability and frequency of failure of a bridge-structure system (Example 5.24).

By the successive addition of odd and even terms, increasingly closer upper and lower bounds can obtained. It should be noted that successive bounds on probability of failure converge, i.e., the ith upper bound is less than the $i - 1$th upper bound and ith lower bound is greater than the $i - 1$th lower bound. However, so far as frequency is concerned this convergence is generally true, but it cannot be guaranteed [13].

Example 5.24 Consider again the bridge system. If we let the failure and repair rates of all components equal 1 per year and 20 per year, respectively, we can plot the lower and upper bounds of probability and frequency of failure as shown in Figure 5.33

An approximate version of min-cut set method

Once the min cut sets have been arranged in series and parallel blocks, series parallel reduction can be done by successively using the formulas for two components in series and parallel. These formulas can be summarized below:

1. First-Order Cut Set k: One component involved:

$$\lambda_{\bar{C}_k} = \lambda_i \tag{5.49}$$

$$r_{\bar{C}_k} = r_i, \tag{5.50}$$

where λ_i, r_i are failure rate and mean duration of component i, and $\lambda_{\bar{C}_k}, r_{\bar{C}_k}$ are failure rate and mean duration of cut set k that contains component i.

2. Second-Order Cut Set k: Two components involved:

$$\lambda_{\bar{C}_k} = \frac{\lambda_i \lambda_j (r_i + r_j)}{1 + \lambda_i r_i + \lambda_j r_j} \tag{5.51}$$

$$r_{\bar{C}_k} = \frac{r_i r_j}{r_i + r_j}, \tag{5.52}$$

where λ_i and λ_j are failure rates of components i and j comprising cut set k, and r_i and r_j are mean failure duration of components i and j comprising cut set k.

A special situation arises when the parallel components are exposed to weather that fluctuates between normal and adverse conditions. In this situation the failure rate in the adverse weather is much higher than in the normal weather. If an average failure rate is used, it can be easily shown that the probability of failure so computed is less than the actual value. There are two ways of dealing with this situation. One is to construct a Markov model of two components in parallel and exposed to the two-state weather. The other is to use approximate equations. The approximate equations are described below:

λ_i, λ'_i = Failure rates of component i in the normal and adverse weather
N, S = Mean duration of normal and adverse weather

$$\lambda_{\bar{C}_k} = \lambda_a + \lambda_b + \lambda_c + \lambda_d \tag{5.53}$$

$$\lambda_a = \frac{N^2}{N+S} \frac{\lambda_i \lambda_j r_i}{N+r_i} \tag{5.54}$$

$$\lambda_b = \frac{S}{N+S} \frac{\lambda_i \lambda_j' r_i^2}{S+r_i} \tag{5.55}$$

$$\lambda_c = \frac{NS}{N+S} \frac{\lambda_i' \lambda_j r_i}{N+r_i} \tag{5.56}$$

$$\lambda_d = \frac{S^2}{N+S} \frac{\lambda_i' \lambda_j' r_i}{S+r_i}, \tag{5.57}$$

where

λ_a = Component failure rate due to both failures occurring during normal weather
λ_b = Initial failure in normal weather, second failure in adverse weather
λ_c = Initial failure in adverse weather, second failure in normal weather
λ_d = Both failures during adverse weather.

It can be seen that the cut-set method focuses on the combination of components that makes the system fail. The following section discusses the tie-set method, which focuses on the combination of components that makes the system work.

5.5.2 Tie-Set Method

We first start with some definitions of tie set and minimal (min) tie set.

Definition 5.3 *A tie set is set of components whose working status will lead to system success.*

Definition 5.4 *A minimal tie set, denoted T, is the a tie set that has no proper subset that is a tie set except itself.*

The min tie set contains a combination of components whose working status makes the system be in success state. Similar to the cut-set method, the term component is used here in a general sense. We can interpret it as a physical component or existence of a condition.

Example 5.25 Find the min tie sets for the system in Figure 5.27.

For this system, the minimum tie sets are $T_1 = \{1, 3\}, T_2 = \{2, 4\}, T_3 = \{1, 4, 5\}$ and $T_4 = \{2, 3, 5\}$.

We can now compute the reliability indices when all the min tie sets are identified using either direct method or the block diagram representation method, which are explained next.

Reliability Indices Calculation Using Direct Method

Recall that each min tie set represents an event where the working status of all components in the set makes a system works. This means that the success of the system is caused by the success of all components in any min tie set.

Let us denote the success event of all components in min tie set T_i, Y is the failure event of the system and $S \setminus Y$ is the success event of the system. We can write:

$$S \setminus Y = T_1 \cup T_2 \cup T_3 \ldots \cup T_n.$$

The probability of success can be simply computed using the expression for the union of events from Chapter 2 as follows:

$$p_s = P(T_1 \cup T_2 \cup T_3 \ldots \cup T_n) \tag{5.58}$$

$$= \sum_i P(T_i) - \sum_{i<j} P(T_i \cap T_j) \tag{5.59}$$

$$+ \sum_{i<j<k} P(T_i \cap T_j \cap T_k) - \ldots \tag{5.60}$$

$$+ (-1)^{n-1} P(T_1 \cap T_2 \cap \ldots \cap T_n). \tag{5.61}$$

Using the conversion rules from Chapter 4, the probability expressions given in (5.61) can be easily converted into frequency of failure. Basically, each term in the probability expression is multiplied by the summation of failure rates of components that are working in that term. Thus, frequency of failure is given by (5.62):

$$f_f = \sum_{i=1}^{n} P(T_i)\bar{\lambda}_i - \sum_{i<j}^{n} P(T_i \cap T_j)\bar{\lambda}_{i+j} \tag{5.62}$$

$$+ \sum_{i<j<k}^{n} P(T_i \cap T_j \cap T_k)\bar{\lambda}_{i+j+k} - \cdots \tag{5.63}$$

$$+ (-1)^n P(T_1 \cap T_2 \cap T_3 \cap T_4 \cap \ldots \cap T_n)\bar{\lambda}_{i+j+\ldots+n}, \tag{5.64}$$

where $\bar{\lambda}_i$ is summation of λ_i of all components $\in T_i$, and

$$\bar{\lambda}_{i+j+\ldots+n} = \sum_{m \in T_i \cup T_j \cup T_k \cup \ldots \cup T_n} \lambda_m. \tag{5.65}$$

Other reliability indices such as mean duration, can be computed from the probability and frequency of failure.

Reliability Indices Calculation Using Block Diagrams

We can find equivalent reliability block diagram using the tie-set method. Since all components in a min tie set need to work to make the system succeed, the components in a tie set can be regarded as connected in series structure. As any tie set can cause system success; the min tie sets are connected in parallel structure. We illustrate this concept using the following example.

Example 5.26 Consider a bridge system shown in Figure 5.27. The min tie sets of this system are found in Example 5.25, which are {1,3}, {2,4}, {1,4,5} and {2,3,5}. We can draw an equivalent reliability block diagram from these min tie sets as seen in Figure 5.34.

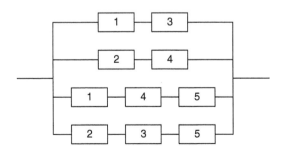

Figure 5.34 Equivalent reliability block diagram of a bridge-structure system using tie-set method.

The probability of success can be calculated from combining the blocks in series and parallel structure. However, similar to the cut-set method, we need to correct the probability expression by setting all exponents to one, which will give the correct probability value.

The frequency of failure can be found using a concept from Chapter 4; we have

$$f_f = f_{T_1 \cup T_2 \cup T_3 \ldots \cup T_n} \tag{5.66}$$

$$= \sum_i f_{T_i} - \sum_{i<j} f_{T_i \cap T_j} \tag{5.67}$$

$$+ \sum_{i<j<k} f_{T_i \cap T_j \cap T_k} - \cdots \tag{5.68}$$

$$+ (-1)^{n-1} f_{T_1 \cap T_2 \cap \ldots \cap T_n}, \tag{5.69}$$

where f_T is the frequency of failure as a result of tie set T.

From the above expression, we need to find the frequency of failure of tie sets T_i, $T_i \cap T_j$, $T_i \cap T_j \cap T_k$, and so on. When all components in the tie set are connected in series, we can easily find the frequency of failure by multiplying the probability of success of each cut set with its equivalent failure rate. From the network reduction method, we know that the equivalent failure rate is simply found from summation of the failure rate of all components. We can now easily calculate the frequency of failure as follows:

$$f_f = \sum_{i=1}^{n} P(T_i)\bar{\lambda}_i - \sum_{i<j}^{n} P(T_i \cap T_j)\bar{\lambda}_{i+j} \tag{5.70}$$

$$+ \sum_{i<j<k}^{n} P(T_i \cap T_j \cap T_k)\bar{\lambda}_{i+j+k} - \cdots \tag{5.71}$$

$$+ (-1)^n P(T_1 \cap T_2 \cap T_3 \cap T_4 \cap \ldots \cap T_n)\bar{\lambda}_{i+j+\ldots+n}. \tag{5.72}$$

where $\bar{\lambda}_i$ is the equivalent failure rate found from the summation of λ_i of all components $\in T_i$, and

$$\bar{\lambda}_{i+j+\ldots+n} = \sum_{m \in T_i \cup T_j \cup T_k \cup \ldots \cup T_n} \lambda_m. \tag{5.73}$$

We arrive at the same formula as found from direct method using conversion rules.

Bounds of Reliability Indices

When the system contains several min tie set, it may become impractical to calculate the exact probability value. We can approximate the probability of success and frequency of failure using the following upper and lower bounds:

p_U = First upper bound to p_f

$$= \sum_i P(T_i)$$

p_L = First lower bound to p_f

$$= p_U - \sum_{i<j} P(T_i \cap T_j)$$

f_U = First upper bound to f_f

$$= \sum_i P(T_i)\lambda_i$$

f_L = First lower bound to f_f

$$= f_U - \sum_{i<j} P(T_i \cap T_j)\lambda_{i+j}.$$

We apply this bounds to the same bridge system as shown below.

Example 5.27   If we let the failure and repair rates of all components equal to 1 per year and 20 per year, respectively, which are the same as in Example 5.24, the lower and upper bounds of probability of success and frequency of failure are shown in Figure 5.35.

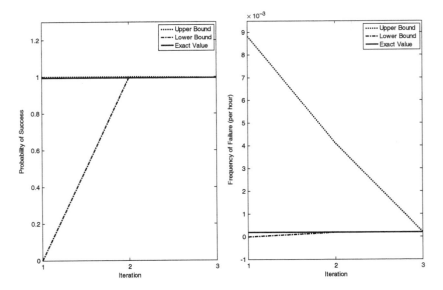

Figure 5.35 Lower and upper bounds of probability of success and frequency of failure of a bridge-structure system (Example 5.27).

5.5.3 Comparison between Cut-Set and Tie-Set Methods

We can see that both cut-set and tie-set methods can be used to calculate reliability indices. It should be noted, however, that in most systems the number of higher probability failure states are less than success states, which makes it comparatively easier to identify the failure states. For this reason, the cut-set method is more widely used in several applications, including power systems.

Another important point to observe is the accuracy of the indices. In most reliable systems the probability of success is well above 90%. When we use the tie-set method to calculate probability of success and its bounds, we may find that the accuracy of the indices from each order of the bound may be insignificant. For example, we can compare the probability of success 0.999 with 0.998—the two probabilities have very small percentage differences. However, when we compare the probability of failure 0.001 with 0.002, we can see that the later is twice as much. This means that using cut-set method will allow us better control over the accuracy of failure probability, especially for a highly reliable system.

Exercises

5.1 Consider a system of two components. Each component can assume two statuses: failure or success. A failure of either component will cause a system to fail. Each component has a failure rate of $\lambda_1 = \lambda_2 = 0.1$ per year and a repair rate of of $\mu_1 = \mu_2 = 10$ per year. In a rare occasion, both components at success state will fail at the same time: assume the common mode failure rate of $\lambda_c = 0.01$ per year. Draw the state-transition diagram of the system and identify failure states. Find frequency of failure of the system. What the are equivalent failure and repair rate of the system?

5.2 Consider a network of seven components shown in Figure 5.36. Use the failure rate of two per year and mean repair time of 24 hours. Using conditional probability on complicating component 7, find system frequency of failure.

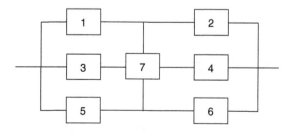

Figure 5.36 A seven-component system.

6

Monte Carlo Simulation

6.1 Introduction

Monte Carlo simulation consists of imitating the stochastic behavior of a physical system. The family of methods that fit this description has been named after the city of Monte Carlo in Monaco because of the resemblance of these methods to games of chance. In general, Monte Carlo simulation methods can be used to imitate any system that exhibits any form of random behavior. In the context of reliability evaluation of a system, Monte Carlo simulation is used to determine how random failures of the components constituting the system affect the reliability of the system.

Monte Carlo simulation is often used as an alternative to analytical methods. Analytical approaches are efficient and should always be employed when it is possible to develop models that are reasonable representations of the physical systems, and also when such models are amenable to solution. Some problems are, however, too complex to be solved in this manner and simulation techniques have to be used.

In simulation, the system is divided into elements whose behavior can be predicted either deterministically or by probability distributions. These elements are then combined to determine the system reliability. Simulation, therefore, also employs a mathematical model, but it proceeds by performing sampling experiments on this mathematical model. Simulation experiments are virtually the same as ordinary statistical experiments, except that they are performed on the mathematical model rather than on the actual system.

The mathematical methods discussed in the previous chapters generally give exact results under the assumptions made, except for numerical round-off. Simulation techniques, however, provide only estimates of the exact results. Moreover they provide only a numerical value, and the whole simulation experiment may have to be repeated to obtain results for even slightly different data. Sensitivity analysis, using a simulation approach, could therefore be quite

Electric Power Grid Reliability Evaluation: Models and Methods, First Edition. Chanan Singh, Panida Jirutitijaroen, and Joydeep Mitra.
© 2019 by The Institute of Electrical and Electronic Engineers, Inc. Published 2019 by John Wiley & Sons, Inc.

computationally expensive. It is, however, a very flexible approach for large and complex systems.

The Concept of Monte Carlo Simulation

Basic concepts of Monte Carlo simulation applied to power systems are described using an example of a system with two independent components. Methodology for random number generation is also given in this section since the simulation relies heavily on these numbers to sample system states.

In simulation, estimation of indices is done in the same manner as in real systems except that a mathematical model of the system, rather than the physical system itself, is used for generating the historical data. This is illustrated with the help of an example of two independent components in parallel. The system is considered failed when both the components are failed. The simulation of this system is conducted by making a mathematical model where the behavior of the components is represented by probability distributions.

Assume that in a system of two components, component 1 is in the up state at the beginning of the observation. Using a random number and the probability distribution of the up time of component 1, the time at which this component will fail is determined. Methods for doing this are explained later in this chapter. In a similar manner, a possible duration of the repair time is generated. A history of the component generated in this manner is one possible realization of the stochastic process. The realization of component 2 is also constructed, and the overlapping outage durations represent the durations of system failure, assuming both components are needed for system success. A number of realizations of the system history can be constructed in this manner and the reliability measures can be obtained from these realizations using statistical methods. In essence, simulation consists of constructing realizations of the stochastic process underlying the system and then extracting the required system performance parameters from these realizations. Most of the refinements in the theory of simulation are concerned either with developing more efficient methods of constructing realizations or extracting the information from the least possible number of realizations.

6.2 Random Number Generation

Random numbers are needed to generate observations from the probability distributions. The basic requirement for the numbers to be random is that they fit a uniform distribution. This means that in a sequence each number should have equal probability of taking on any one of the possible values and it must be statistically independent of the other numbers in the sequence. There are several methods available to generate random numbers, and only one is described here.

It is a multiplicative congruential method and obtains the $(n + 1)$th random number R_{n+1} from the nth random number R_n by using the following recurrence relation due to Lehmer:

$$R_{n+1} \equiv (aR_n \bmod m), \tag{6.1}$$

where a and m are positive integers, $a < m$. The above notation signifies that R_{n+1} is the remainder when (aR_n) is divided by m. The first random number or the seed, R_0, is assumed, and the subsequent random numbers can be generated by this recurrence relation. The sequence of random numbers so generated has a cycle starting when the generated random number is again equal to the seed. Great care has to be exercised in the selection of a combination of R_0, a and m. The sequence cycle should be larger than the number of random numbers required. One satisfactory combination is

$a = 455470314$
$m = 2^{31} - 1 = 2147483647$
$R_0 =$ any integer between 1 and 2147483646.

Now if random numbers between, say, 0 and 999 are required, then the computer can be instructed to take the last three digits of the random number generated. It can be seen that the sequence produced is predictable and reproducible and is not therefore strictly a sequence of random numbers. For this reason these numbers are called pseudorandom numbers. They can, however, satisfactorily play the role of random numbers in a computer simulation, as long as the simulation completes before the sequence begins to repeat. In fact, in many applications where alternative design configurations are being evaluated, the use of the same sequence of random numbers may be desirable.

6.3 Classification of Monte Carlo Simulation Methods

Most Monte Carlo simulation methods used for system reliability analysis can be classified as sequential or nonsequential. Random sampling, or nonsequential simulation, consists of performing random sampling over the aggregate of all possible states the system can assume during the period of interest. In sequential methods, on the other hand, the mathematical model of the system is made to generate an *artificial history* over time, and appropriate statistical inferences are drawn from *this history*. After an appropriate number of samples have been drawn from the aggregate population, sample statistics are used to estimate the corresponding population statistics, which are discussed in this chapter. The Monte Carlo simulation methods are described below.

6.3.1 Random Sampling

Non sequential simulation proceeds by performing proportionate sampling on the state space. Proportionate sampling is a method of performing random sampling such that the probability of drawing a particular system state equals the actual probability of the physical system assuming that state. A sufficient number of states are drawn to construct a sample set of states that can be assumed, with an acceptable degree of confidence, to be representative of the population of states in the aggregate state space. Statistical estimates of the system reliability are determined from the sample set of system states.

The basic step in system state sampling is the sampling of component states using random numbers uniformly distributed between zero and one. The component states then determine the system state. Two basic approaches to component state sampling are described as follows.

Proportional Probability Method
In this approach each state is assigned a range between zero and one proportional to the probability of occurrence of that state. If the random number drawn falls in that range, the state is assumed to occur. In this fashion the states are sampled proportional to their probability. The process can be understood by an example of a component that has the probability distribution described in Table 6.1.

The random number range can now be assigned as shown in Table 6.2. Once the random number is drawn, it can be seen which state is sampled. For example, if the random number is between 0.3+ and 0.7, then state 3 is sampled. It is easy to see that by this process, the states are sampled proportional to their probabilities.

Probability Distribution Method
The state sampling can also be performed using the cumulative probability distribution function. This can be understood by taking the example described earlier. The probability distribution is given in Table 6.1 and the probability mass function of this component is shown in Figure 6.1. Next to the mass

Table 6.1 Example Component Probability Distribution

State (random variable)	Probability
1	0.1
2	0.2
3	0.4
4	0.2
5	0.1

Table 6.2 Assignment of Range to States

Random number drawn	State sampled
0 to 0.1	1
0.1+ to 0.3	2
0.3+ to 0.7	3
0.7+ to 0.9	4
0.9+ to 1.0	5

function is shown the cumulative distribution function. Now the state can be simply sampled by inserting the random number drawn on the vertical axis and reading the sampled state on the horizontal axis. It is easy to see that this process basically accomplishes the same objective as the proportional allocation but is more convenient and easier to implement.

When system consists of n independent components, then to sample a system state n random numbers are needed to sample the state of each component. Let us consider a system consisting of two components with the probability distribution shown in Figure 6.1. For this system, sampling may proceed as shown in Table 6.3.

Now let us say that we want to estimate the probability of state (3,3). This can be done as follows:

$$Pr(3, 3) = \frac{n}{N}, \tag{6.2}$$

where N is the total number of samples and n is the number of samples with state $(3, 3)$.

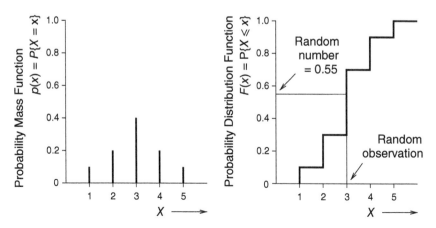

Figure 6.1 Sampling from a probability distribution.

Table 6.3 System State Sampling for Two-Component Systems

RN for component 1	RN for component 2	System state
0.946	0.601	(5,3)
0.655	0.671	(3,3)
0.791	0.333	(4,3)
0.345	0.532	(3,3)
0.438	0.087	(3,1)
0.311	0.693	(3,3)
0.333	0.918	(3,5)
0.998	0.209	(5,2)
0.923	0.883	(5,4)
0.851	0.135	(4,2)
0.651	0.034	(3,1)
0.316	0.525	(3,3)
0.965	0.427	(5,3)
0.839	0.434	(4,3)

From Table 6.3, $Pr(3,3) = \frac{4}{14} = 0.286$, and the actual probability of (3,3) is $0.4 \times 0.4 = 0.16$. The difference between the estimated value and the true value underscores the importance of sample size. If this sampling and estimation are continued, the estimated value will approach 0.16 with an appropriate sample size. Therefore, one must have some criterion to decide whether the sample size is appropriate and the indices have converged. Convergence means that the estimated value is within the desired accuracy with the specified probability. Estimation and convergence are discussed in Section 6.5. The sequential simulation method is described next.

6.3.2 Sequential Sampling

Sequential simulation methods can be classified into two categories; namely, Fixed-time interval method and Next event method, depending upon how the time is stepped through. A brief description of each is given below.

Fixed-Time Interval Method

The basic time interval Δt is chosen depending upon the operating characteristics of the system. Starting in the initial state, time is advanced by Δt, and it is determined if an event has occurred. If no event has occurred then the system

Table 6.4 Fixed-Time Interval Example: Initial State 0

RN	Event
0 to 0.3	stay in 0
0.3 to 1.0	transit to 1

stays in the same state; otherwise the system state is changed depending on the event. These two steps may be repeated to get statistical convergence of the indices being calculated. This method is useful for simulating Markov chains, where transition probabilities are defined over a time step. The simulation is performed using basically the proportionate allocation technique described under sampling. This is indeed sampling conditioned on a given state, as can be seen from the following example. Assume that there is a two-state system, and the probability of transiting from state to state over a single step is given by the following matrix:

$$
\begin{array}{cc}
 & \text{Final state} \\
\text{Initial state} & \begin{array}{cc} 0 & 1 \end{array} \\
\begin{array}{c} 0 \\ 1 \end{array} & \begin{bmatrix} 0.3 & 0.7 \\ 0.4 & 0.6 \end{bmatrix}.
\end{array}
\tag{6.3}
$$

The realizations are constructed using a table of random numbers. Starting in state 0, a random number is drawn, and the next state is determined using Table 6.4. Similarly, if the process is in state 1, the next state is determined using Table 6.5. Realization over 10 steps is shown in Table 6.6.

Next Event Method

This process proceeds by keeping a record of the next simulated events scheduled to occur. The most imminent event is assumed to have occurred, and the simulated time is advanced to the point of occurrence of the event. The cycle is repeated as many times as needed for convergence. This type of simulation corresponds to the continuous-time Markov processes, and here

Table 6.5 Fixed-Time Interval Example: Initial State 1

RN	Event
0 to 0.4	transit to 0
0.4 to 1.0	stay in 1

Table 6.6 Fixed-Time Interval Example: Realization

Step	RN	State
1	0.947	0
2	0.601	1
3	0.655	1
4	0.671	1
5	0.791	1
6	0.333	1
7	0.345	0
8	0.531	1
9	0.478	1
10	0.087	1

the state-residence times of components are defined by distributions of continuous random variables. Therefore an essential step here is the determination of the value of a random variable from the continuous distribution.

The procedure described as the fixed-time interval method in section 6.3.2 can also be used for continuous distributions, where continuous distributions are approximated by discrete distributions whose irregularly spaced points have equal probabilities. The accuracy can be increased by increasing the number of intervals into which (0,1) is divided. This requires additional data in the form of tables. Although the method is quite general, its disadvantages are the great amount of work required to develop tables and possible computer storage problems. The following analytic inversion approach is simpler. Let z be a random number in the range zero to one with either a uniform probability density function or a triangular distribution function, i.e.,

$$f(z) = \begin{cases} 0 & Z < 0 \\ 1 & 0 \leq Z \leq 1. \\ 0 & 1 < Z \end{cases} \tag{6.4}$$

Similarly,

$$F(z) = \begin{cases} 0 & Z < 0 \\ z & 0 \leq Z \leq 1. \\ 0 & 1 < Z \end{cases} \tag{6.5}$$

Let $F(x)$ be the distribution function from which the random observations are to be generated, and $z = F(x)$. Solving the equation for x gives a random observation of X. Thus, the generated observations have $F(x)$ as the probability distribution. This can be shown as follows.

Let φ be the inverse of F, then

$$x = \varphi(z). \tag{6.6}$$

Now x is the random observation generated. To determine its probability distribution,

$$Pr(x \leq X) = Pr(F(x) \leq F(X)) = Pr(z \leq F(X)) = F(X). \tag{6.7}$$

Therefore the distribution function of x is $F(x)$ as required. In the case of several important distributions, special techniques have been developed for efficient random sampling. Here only the exponential distribution is discussed.

The exponential distribution has the following probability distribution in (6.8):

$$P(X \leq x) = 1 - e^{-\rho x}, \tag{6.8}$$

where $\frac{1}{\rho}$ is the mean of the random variable X.

Setting this function equal to a random decimal number between zero and one,

$$z = 1 - e^{-\rho x}. \tag{6.9}$$

Since the complement of such a random number is also a random number, the above equation can be written as (6.10):

$$z = e^{-\rho x}. \tag{6.10}$$

Take the natural logarithm of both sides and we have

$$x = -\frac{\ln(z)}{\rho}, \tag{6.11}$$

which is the desired random observation from the exponential distribution having $\frac{1}{\rho}$ as the mean.

As an example of a two-state component, the next event method can be illustrated by taking an example of two components whose up and down times are exponentially distributed. The failure and repair rates of these components are given in Table 6.7. The simulation of these two components is shown in Table 6.8.

In Table 6.8, "←" indicates the sampled time causing change and is added to obtain the total time. For example at time 0, two random numbers are drawn

Table 6.7 Next Event Method Example

Component	λ (f/hr)	μ (rep/hr)
1	0.01	0.1
2	0.005	0.1

for the time to failure of components 1 and 2. These numbers are substituted in (6.11), which yields 5 and 101 hours respectively for components 1 and 2. Since 5 is smaller, component 1 fails leading to state DU (component 1 is down and component 2 is up) at hour 5. At hour 5, component 1 has no time left in success state, while the component 2 has 96 hours left to fail. A random number is then drawn for component 1 to find its time to repair using (6.11), which yields 4 hours. The process proceeds in this fashion, as shown in Table 6.8.

6.4 Estimation and Convergence in Sampling

It is crucial to sample sufficient number of states to estimate reliability indices. This section describes the estimation and convergence criterion of both

Table 6.8 Simulation of Two Components by the Next Event Method

Time	Random Number for Component		Time to change		Component State*	
	1	2	1	2	1	2
0	0.946	0.601	5 ←	101	U	U
5	0.655	-	0/4 ←	96	D	U
9	0.670	-	0/40 ←	92	U	U
49	0.790	-	0/2 ←	52	D	U
51	0.332	-	0/110	50 ←	U	U
101	-	0.345	60	0/11 ←	U	D
112	-	0.531	49 ←	0/127	U	U
161	0.437	-	0/8 ←	78	D	U
169	0.087	-	0/244	70 ←	U	U
239	-	0.311	174	0/12 ←	U	D
251	-	0.693	162	0/73 ←	U	U
324	-	0.333	89	0/	U	D
		↓				

*D indicates Down States and U indicaties Up states.

techniques, namely: random sampling and sequential sampling. These methods are described below.

6.4.1 Random Sampling

Let $x = (x_1, x_2, \ldots, x_m)$ be the state of power system, where

x_i = State of ith component
X = Set of all possible system states
$P(x)$ = Probability of state x.

Let $F(x)$ be the test applied to verify if state x is able to satisfy the load. The expected value of $F(x)$ is given in (6.12):

$$E[F(x)] = \sum_{x \in X} F(x)P(x). \tag{6.12}$$

For $E[F(x)]$ to be *LOLP*,

$$F(x) = \begin{cases} 1 & \text{if loss of load} \\ 0 & \text{otherwise} \end{cases}. \tag{6.13}$$

In random sampling, $x \in X$ are sampled from their joint distributions. Then the estimate of $E[F(x)]$ is (6.14):

$$\hat{E}[F(x)] = \frac{1}{N_S} \sum_{i=1}^{N_S} F(x_i), \tag{6.14}$$

where

N_S = Number of samples
$F(x_i)$ = Test result for ith sampled value,

and the variance of this estimate is

$$Var(\hat{E}[F(x)]) = \frac{Var(F(x))}{N_S}. \tag{6.15}$$

Since $Var(F(x))$ is not known, its estimate can be used as in (6.16):

$$\hat{Var}(F(x)) = \frac{1}{N_S} \sum_{i=1}^{N_S} (F(x_i) - \hat{E}[F(x)])^2. \tag{6.16}$$

Convergence is typically based on the value of the coefficient of variation (COV), which is given in (6.17):

$$COV = \frac{SD(\hat{E}[F(x)])}{\hat{E}[F(x)]}, \tag{6.17}$$

where

$SD(\hat{E}[F(x)]) =$ Standard deviation of the estimator of $E[F(x)]$.

Since $SD(\hat{E}[F(x)]) = \sqrt{Var(\hat{E}[F(x)])}$, using (6.16) we have

$$COV = \frac{1}{\hat{E}[F(x)]} \sqrt{\frac{Var(F(x))}{N_S}}. \tag{6.18}$$

Hence,

$$N_S = \frac{Var(F(x)}{(COV \times \hat{E}[F(x)])^2}. \tag{6.19}$$

It can be seen from equation (6.19) that

1. Sample size is not affected by system size or complexity.
2. Accuracy required and the probability being estimated effect the sample size.
3. Computational effort depends on number of samples and CPU time per sample.

6.4.2 Sequential Sampling

In general terms, states of the system are generated sequentially by transition from one state to the next using probability distributions of component state durations and random numbers from [0,1].

Consider a component i; assume that this component is up and duration of up state is given by U_i (a random variable). If Z is a random number, then the observation of up time can be drawn by using (6.20):

$$z = Pr(U_i \leq U) = F_i(U). \tag{6.20}$$

A component with minimum time makes a transition and causes system transition. Sampling time to transition is done in a fashion similar to random sampling, as shown in Figure 6.2. This is usually determined analytically, as described in Section 6.3.2.

The algorithm can be described in the following steps. Let us assume that the nth transition has just taken place at time t_n, and the time to next transition of component i is given by T_i. Thus the vector of times to component transitions is given by T_i, and the simulation proceeds in the following steps:

Step 1. The time to next system transition is given by (6.21):

$$T = \min T_i. \tag{6.21}$$

If this T corresponds to T_p, that is, the pth component, then next transition takes place by the change of the state of this component.

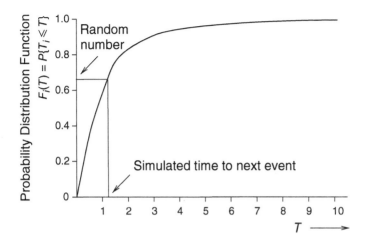

Figure 6.2 Simulating time to next event.

Step 2. The simulation time is now advanced:

$$t_{n+1} = t_n + T. \tag{6.22}$$

Step 3. The residual times to component transitions are calculated by (6.23):

$$T_i^r = T_i - T, \tag{6.23}$$

where $T_i^r =$ Residual time to transition of component i.

Step 4. The residual time for component p causing transition becomes zero, and time to its next transition T_p is determined by drawing a random number.

Step 5. The time T_i is set as shown in (6.24):

$$T_i = \begin{cases} T_i^r & i \neq p \\ T_p & i = p \end{cases}. \tag{6.24}$$

Step 6. From t_n to t_{n+1}, the status of equipment stays fixed, and the following steps are performed:

1. The load for each node is updated to the current hour.
2. If no node has loss of load, the simulation proceeds to the next hour, otherwise state evaluation module is called.
3. If, after remedial action, all loads are satisfied, then simulation proceeds to next hour. Otherwise, this is counted as loss of load hour for those nodes and the system. Also if in the previous hour there was no loss of load, then this is counted as one event of loss of load.
4. Steps 1–3 are performed until t_{n+1}.

Step 7. The statistics are updated as described after Step 8, and the process moves to Step 2.

Step 8. The simulation is continued until convergence criterion is satisfied. Let

I_i = Value of reliability index (for example number of hours of load loss) obtained from simulation data for year i

N_y = Number of years of simulated data available

SD_I = Standard deviation of the estimate I

Then, estimate of the expected value of the index I is given in (6.25), and standard deviation is given in (6.26):

$$\hat{I} = \frac{1}{N_y} \sum_{i=1}^{N_y} I_i \qquad (6.25)$$

$$SD_I = \sqrt{\frac{\sum_{i=1}^{N_y}(I_i - \bar{I})^2}{N_y}}. \qquad (6.26)$$

Note that SD_I, the standard deviation of the estimate, $\bar{I}$, varies as $\frac{1}{\sqrt{N_y}}$ and will approach zero as N_y approaches infinity.

6.5 Variance Reduction Techniques

It has been pointed out that the precision of sample estimates can be increased by making the sample size large enough. Increasing the precision is equivalent to decreasing the variance of the sample estimates. The simple method of repeated runs (or making a single run very large and dividing it into equal intervals), treating the measures obtained from each sample as independent sample values until the variance has been reduced to the desired level, is usually quite time-consuming. Special techniques for reducing variance have been devised. These techniques extract as much and as precise information as possible from the amount of simulation that can be economically executed. There are several variance reduction techniques, such as importance sampling, control variates, antithetic variables and stratification.

Importance sampling modifies the distribution functions of system components so that loss of load states occur more frequently. Control variates and antithetic variable methods manipulate variance of the reliability indices by introducing a new random variable that has the same mean value yet lower variance. The methods achieve variance reduction by exploiting correlations between the new random variable and the actual index. These methods alter the structure of actual random variables by either modifying distributions or the random variable itself. The obtained distributions of the estimates are not as representative of the distributions of the actual indices and may be used only to estimate the mean value.

In stratification, for example, proportional sampling groups states into mutually exclusive strata before sampling to improve the representativeness of the samples. This method may not be practical for large system analysis since it may be difficult to select criteria for stratification. Another variation, a mixed technique between random sampling and stratified sampling, is Latin Hypercube sampling (LHS). This technique was developed to assess uncertainties for a complex system with multirandom variables. LHS can estimate both mean value and distribution of the indices. The method, however, is computationally expensive due to complete construction of the equivalent probability function and space required during sampling. The main disadvantage of this method is that the sample sizes need to be determined in advance of the sampling. The reader is referred to [14, 15] for further discussion of these methods.

6.5.1 Importance Sampling

The idea of importance sampling can be explained by (6.12) and (6.13) for calculating LOLP. Equation (6.12) is an analytical expression for calculating the LOLP and in (6.13), the estimate of LOLP, $\hat{E}[F(x)]$ is obtained by sampling states using the probability mass function $P(x)$ of the random variable X. In importance sampling the idea is to accelerate the process by sampling from a different probability mass function, say $g(x)$, such that the variance of the estimate is reduced leading to a faster convergence. Now equation (6.12) can be rewritten as:

$$E[F(x)] = \sum_{x \in X} F(x)P(x) \tag{6.27}$$

$$= \sum_{x \in X} F(x)g(x)\frac{P(x)}{g(x)} \tag{6.28}$$

$$= E_g[F(x)\frac{P(x)}{g(x)}], \tag{6.29}$$

where E_g is the expected value with respect to the probability mass function $g(x)$, and the term $\frac{P(x)}{g(x)}$ is called the likelihood ratio.

In terms of non sequential simulation, (6.29) can be written as the estimate with sampling done from importance sampling distribution $g(x)$ as follows:

$$\hat{E}(F(x)) = \frac{1}{N_s}\sum_{i=1}^{N_s} F(x_i)\frac{P(x)}{g(x)}. \tag{6.30}$$

The objective is to select an optimal sampling density or mass function $g(x)$ such that the sample variance is minimized. It has been shown that the optimal solution is to use the following density function for importance sampling:

$$g^{opt}(x) = \frac{F(x)P(x)}{E(x)}. \tag{6.31}$$

In this manner only one sample will need to be drawn, as can be seen by substituting $g^{opt}(x)$ for $g(x)$ in (6.30). It is obvious that this is not of any value since we will need to know $E(x)$ for this in the first place, and then there is no reason to perform the Monte Carlo simulation. A sensible approach would be to use a sampling distribution with minimum divergence from the optimal sampling distribution. An effective approach for a certain class of distributions is to find the parameters in the sampling distribution such that the Kullback-Leibler divergence between g^{opt} and the one actually used, g, is minimized. The Kullback-Leibler divergence between two densities $g(x)$ and $h(x)$ is given by:

$$D(g, h) = E_g[\ln g(x) - \ln h(x)], \tag{6.32}$$

where the subscript g indicates that expectation is with respect to $g(x)$. It has been shown that for several distributions the problem can be solved using stochastic minimization. In this case of power systems, if the components are assumed independent, and the unavailability of component j is given by u_j, then the probability distribution can be defined by binomial distribution of state variable x. To use importance sampling, the unavailability can be altered to v_j such that the computation time is minimized. The solution to this class of problems has been found in [16] and used, for example, in [17]. For example, consider n_g generators that are independent, with u_j as the unavailability of jth generator and v_j as the altered unavailability of the jth generator. The likelihood ratio is the correction made in the sampling process to normalize the probabilities as the sampling is being done from the altered v_j. The likelihood ratio for the state of the generation system can be expressed as

$$W(X_i, u, v) = \frac{\prod_{j=1}^{n_g} (u_j)^{1-s_j}(1 - u_j)^{s_j}}{\prod_{j=1}^{n_g} (v_j)^{1-s_j}(1 - v_j)^{s_j}}. \tag{6.33}$$

Here the jth element of vector u is the original unavailability of the jth generator, the jth element of vector v is the altered unavailability of generator j and s_j is the status of generator j, with $s_j = 0$ if the generator is up and $s_j = 1$ if the generator is down. To find the optimal vector v, some preliminary work will need to be done. An algorithm for obtaining vector v for this problem is described in [17]. It should, however, be remembered that the calculation of the parameter vector for the importance sampling distribution is system dependent, although the concept of using Kullback-Leibler divergence is quite general.

6.5.2 Control Variate Sampling

Let Z be a random variable that is strongly correlated with the random variable F. Define $Y = F - \alpha(Z - E(Z))$. Then,

$$E(Y) = E(F) \tag{6.34}$$
$$V(Y) = V(F) + \alpha^2 V(Z) - 2\alpha \text{Cov}(F, Z). \tag{6.35}$$

Now, if $2\alpha \text{Cov}(F, Z) > \alpha^2 V(Z)$, then

$$V(Y) < V(F). \tag{6.36}$$

Equation (6.36) implies that if the process is made to converge on Y rather than Z, then it will converge faster. At the same time, (6.34) ensures that the mean value that we compute through Y will be that of Z, so the acceleration of convergence will not introduce any bias into the estimate.

6.5.3 Antithetic Variate Sampling

The concept of antithetic variables can be explained as follows. Let

$$F_\alpha = \frac{1}{2}(F' - F'').$$

If $E(F') = E(F'') = E(F)$, then

$$E(F_\alpha) = E(F)$$
$$V(F_\alpha) = \frac{1}{4}\left[V(F') - V(F'')\right] + 2\text{Cov}(F', F'').$$

If F' and F'' are negatively correlated, then $\text{Cov}(F', F'') < 0$, and hence

$$V(F_\alpha) < V(F).$$

Negative correlation can be produced by using random numbers Z_i to compute $E(F')$ and $(1 - Z_i)$ to compute $E(F'')$.

6.5.4 Latin Hypercube Sampling

Latin Hypercube Sampling (LHS), developed by McKay, Conover and Beckman in 1979, combines the concepts of stratified and random sampling [14, 15]. The sample size, n, is pre selected, and the probability distribution function is divided into n intervals with equal probability of occurrence. Random sampling is then performed for each interval corresponding to the probability distribution function in that interval. This means that LHS is a constrained version of Monte Carlo sampling that can produce estimates more precise than Monte Carlo with the same sample size. LHS is thus considered one of the variance reduction techniques. This major advantage can be exploited when combining reliability evaluation with optimization. LHS is currently used in the stochastic optimization framework in comparison with Monte Carlo sampling. The results show that the method produces tighter bounds of the optimal solution than Monte Carlo [18, 19]. In addition, the approximate distributions of the reliability indices are also generated, which may be used as a risk assessment tool since the reliability index itself only represents the mean value. LHS was invented to estimate uncertainty in a problem where the variable of interest is expressed as a function of random variables [14] as follows:

$$y = f(x), \tag{6.37}$$

where y is the variable of interest and x is a vector of random variables.

When the function to be evaluated is complicated and computation intensive, investigation of interaction of variable of interest with other stochastic variables in multi dimensions is inevitably cumbersome. LHS has been developed as a probabilistic risk assessment tool to specifically assist this type of investigation. While Monte Carlo is conventionally applied for power system reliability problems, LHS can yield a relatively better estimate of the distribution of the variable of interest than Monte Carlo [14, 15]. This is due to the fact that prior to sampling, LHS divides a distribution function into intervals of equal probability. The number of intervals is equal to sample size. Then LHS randomly chooses one value from each and every subinterval with respect to its distribution in that subinterval. This means that LHS is done over the entire spectrum of the distribution function, including the tail-end values. The sampled values thus represent the actual distribution better than Monte Carlo, especially for the extreme region of the distribution. The variance of a sample from LHS is considered smaller than that from Monte Carlo since LHS yields a stratified sample of the random variables.

With multi dimensional random variables, LHS creates a system state by pairing the values after sampling each random variable individually. The pairing scheme is rather simple in the case of uncorrelated random variables. A system state is found by randomly choosing one value out of the sampled values from each component without replacement. However, in cases of random variables with some correlation among them, a strategy using a pairing scheme will involve use of optimization. Interested readers are referred to [20] for detailed procedure in the presence of correlations between random variables.

Exercises

6.1 Consider an electric system having two generating units connecting to a bus sending electricity to the load through one transmission line. Assume that each generator can deliver 50 MW output with failure probability of

0.01. The transmission line has 100 MW transfer capability and has failure probability of 0.1. A system load is 50 MW with probability 0.85, and 100 MW with probability 0.15. Assume that all components fail independently; find the system loss of load probability using random sampling when a sample size is 100. Verify your results using probability rules.

6.2 An electric system consists of one generator and two transmission lines. The 100-MW generator supplies a 100 MW load through two transmission lines in parallel. Each transmission line has maximum transfer capability of 100 MW. The failure rate of a generator is 0.01 per day, while its mean repair time is 12 hours. The failure rate of the two transmission lines are 10 and 15 times per year, while their mean repair time are 6 hours. Assume that all components fail and repair independently. Find the loss of load probability using random sampling method when the number of samples is 100, 500 and 1000 states, respectively. Then, compare the simulation results with the exact value from analytical methods.

Part II

Methods of Power System Reliability Modeling and Analysis

7

Introduction to Power System Reliability

7.1 Introduction

In Part I of this text we developed methods that are suitable for reliability analysis of systems consisting of many components. We used knowledge of the reliabilities of individual components, and of the way these components interact, to analyze the reliability of the system that the components constitute. In Chapter 6 we developed the basics of Monte Carlo simulation, which is often used as an alternative to the analytical methods presented. In the following chapters we will present methods of applying or adapting these techniques to determine, through analysis or simulation or a combination of both, the reliability of electric power systems. Owing to the size and complexity of power systems, the reliability evaluation of these systems generally requires the application of specialized techniques, both in modeling power system components and systems as a whole, and in performing the evaluation. Moreover, the industry has developed its own parlance where reliability indices are concerned, and some of these are quite different from the indices that we encountered in Part I of the text. We will describe some of indices and some of these techniques in this Part II.

7.2 Scope of Power System Reliability Studies

Power system reliability studies usually focus on one of the following functional zones in the system:

1. Generation system: This kind of study assesses the adequacy of the generation system, i.e., its ability to meet the total load on the system. This is discussed in detail in Chapter 8.
2. Transmission system: This kind of study evaluates the integrity of the transmission system and its ability to transfer the necessary power from the generation to the load.

Electric Power Grid Reliability Evaluation: Models and Methods, First Edition. Chanan Singh,
Panida Jirutitijaroen, and Joydeep Mitra.
© 2019 by The Institute of Electrical and Electronic Engineers, Inc. Published 2019 by John Wiley & Sons, Inc.

3. Distribution system: This kind of study determines load point reliability indices. This will not be covered in this text, but more information can be found in [21].

4. Interconnected system or multi node system: This kind of study evaluates the reliability of the combined generation and transmission system. There are two classes of multi node system studies: composite system reliability and multi-area reliability. These are discussed in greater detail in Chapters 10 and 11.

5. Protection system: This kind of study assesses the reliability of protection systems and the effect of their failures on the power system. This will not be covered in this text, but more information on definitions and practices can be found in [22, 23], while models and analytical methods are discussed in [24, 25].

6. Industrial and commercial systems: This kind of study focuses on determining means of enhancing reliability and maintaining service in industrial and commercial systems. This will not be covered in this text, but more information can be found in [26].

Due to the size and complexity of power systems, reliability analysis of an "entire power system" is never evaluated, but a set of sequential studies may be performed. The functional zones of interest are evaluated and appropriate indices are determined to fit the scope of the application. For instance, if the reliability of service at a particular load point is required to be evaluated, an interconnected system analysis is performed to determine the reliability of service at the distribution substation that serves that load point, then a reliability assessment is performed for that distribution system, and the results are combined to determine the required load point index.

7.3 Power System Reliability Indices

Power system reliability is concerned with the ability of a power system to satisfy the load on the system while operating the system in a secure manner, i.e., in a manner that respects the capabilities of the generation and transmission (and distribution) components and does not violate these capabilities for extended periods of time. A component of system reliability is system *security*, which deals with the relative ability of the system to survive system disturbances, such as faults or equipment failures, without these disturbances evolving into cascading failures or system instability[1].

---

1 This is the traditional definition of system security. In recent years, the definition of system security has been somewhat altered to denote the system's ability to withstand reliability events resulting from both inadvertent phenomena and malicious attacks.

Power system reliability indices, as well as the evaluative methods used to determine these indices, can be classified into two categories—predictive and empirical. Predictive indices are determined from information pertaining to component reliability and the manner in which components constitute the system—system configuration, behavior, operational characteristics and physical and temporal relationships, if any. The methods described in this text are used for determining predictive indices. Empirical indices, on the other hand, are determined from direct observation, such as by collecting data at the location of interest. For instance, there are predictive methods for reliability evaluation of distribution systems, but in many instances distribution reliability indices are reported from direct collection of data at load points. Formulas and methods for determination of distribution reliability indices are provided in [21].

Another classification exists for power system reliability indices. This classifies indices into the following categories:

1. Probability and expectation indices: These indices include loss of load probability (LOLP) and loss of load expectation (LOLE) indices. LOLE indices may be expressed as hourly loss of load expectation (HLOLE, given in hours per year), which indicates the expected number of hours during a year when the system or parts of the system will experience service failures, or as daily loss of load expectation (DLOLE, given in days per year), which indicates the expected number of days during which the system or parts of the system will experience service failures during peak hours.
2. Frequency and duration indices: These indices indicate the frequency and duration of reliability events. Examples include loss of load frequency (LOLF, expressed in failures per year) and mean time between failures (MTBF, usually expressed in hours). Several distribution reliability indices, such as system average interruption duration index (SAIDI) and system average interruption frequency index (SAIFI) also belong to this category.
3. Severity indices: These indices convey the severity of reliability events. Examples include expected power not served (EPNS, expressed in MW per year) and expected unserved energy (EUE, also known as expected energy not served or EENS, expressed in MWh per year).

The above quantities can be better understood by examining Figure 7.1. This figure shows the capacity and load profile of a typical system over a typical year. It also shows the conditions contributing to deficiency or loss of load and provides a physical interpretation of frequency, duration and severity indices of system reliability.

All the above indices are probabilistic measures of system reliability. In the early days of system operation, when probabilistic indices were not well developed, some deterministic indices were used. These indices essentially pertain to

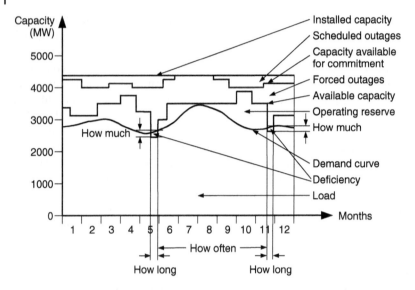

Figure 7.1 Capacity, demand and deficiency profile over one year.

generation adequacy and are still used today in combination with probabilistic indices. The following are the deterministic indices of generation adequacy:

1. Percent Reserve Margin: This is defined as the excess of installed generating capacity over annual peak load expressed in percent of annual peak load. It does not directly reflect system parameters such as unit size, outage rate and the load shape, but it does provide a reasonable relative estimate of reliability performance if parameters other than margin remain essentially constant.
2. Reserve margin in terms of largest unit: The planned reserve is kept equal to or bigger than the size of the largest unit. This index recognizes the importance of unit capacities in relationship to reserve margin; however, it does not recognize all other factors that affect system reliability.

7.4 Considerations in Power System Reliability Evaluation

It should be evident by now that the kind of reliability analysis that one is performing will drive the considerations that go into modeling the components and their interactions and the approach that will be used to assess the system reliability. System size, topology, geographical expanse, diversity of loads, operating policies and statistical and temporal dependencies between components are all considerations that drive the models and methods. It is impossible to generalize how the models and methods are influenced by all these factors, but some of the approaches used will be illustrated in following chapters.

It is also important to define system "failure" in the context of reliability evaluation. Since it is common practice to use loss of load as a criterion, it is appropriate to define the phenomenon that results in a *reliability event* or *interruption*. This phenomenon is the *forced outage* of one or more components and is also known as a *contingency*.

Component outages can be of two types—*scheduled* or *forced*. A *scheduled outage* occurs when a component is taken out of service for a predetermined period of time, usually for the purpose of preventive maintenance. A *forced outage* occurs when a component malfunctions or fails due to random or unforeseen events, such as equipment malfunction or human error. It is this latter class of outages, i.e., *forced outages*, with which probabilistic reliability assessment is concerned. Forced outages can take several forms, but these will not be described in detail here. References [27] and [28] are excellent resources for the reader who wishes to learn more about the different types and classifications of outages.

In Chapters 8 through 11, we will illustrate several approaches to performing reliability modeling and evaluation in different functional zones of a power system. While these approaches are by no means exhaustive, they are illustrative, and the reader will be provided with ample references to other sources that enable the reader to learn more about these methods.

8

Generation Adequacy Evaluation Using Discrete Convolution

8.1 Introduction

This chapter describes the process of generation adequacy evaluation using the method of discrete convolution. In this process, the transmission facilities are assumed capable of transporting the generation to load points. The focus is on the evaluation of adequacy of generation resources. In the next chapter, the effect of the transmission system will be considered. The material covered in this chapter also serves to explain how power system components are modeled for the purpose of reliability analysis.

The method of discrete convolution consists of building generation and load models in the form of probability and frequency distributions of the random variables representing the total system generation and the total system load. The generation model is usually constructed in terms of capacity levels on outage and is known as the capacity outage probability and frequency table (COPAFT). The load model is built as probability and frequency distributions of the total system load. The two models, i.e., the distributions of C_O (capacity outage) and L (load) are used to build the *generation reserve model*, which consists of the probability and frequency distributions of the random variable representing the excess of system generation over load. These distributions are obtained by performing discrete convolution between the generation and load models. The generation adequacy indices are determined from the generation reserve model.

The steps involved in this process are described below.

8.2 Generation Model

In prior chapters we have examined Markov models of system components. Generating units can be modeled as Markovian components with two or more states. A two-state model would consist of an up state and a down state, while

Electric Power Grid Reliability Evaluation: Models and Methods, First Edition. Chanan Singh, Panida Jirutitijaroen, and Joydeep Mitra.

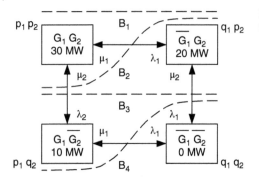

Figure 8.1 System of two two-state units.

models with a larger number of states can consist of up and down states and several intermediate, or *derated*, states. In this chapter we will mostly work with two- and three-state models of generating units; the extension to larger numbers of states is not particularly difficult.

Consider a system consisting of two independent generators: one with capacity 10 MW, failure rate λ_1 and repair rate μ_1, the other with 20 MW, λ_2 and μ_2. The state-transition diagram for this system is as shown in Figure 8.1.

Also shown in Figure 8.1 are the probabilities and capacities of the four system states, and three broken lines labeled B_1, B_2, B_3 and B_4. The probabilities are expressed in terms of $p_1 = P\{G_1\}$, $q_1 = P\{\overline{G_1}\}$, $p_2 = P\{G_2\}$ and $q_2 = P\{\overline{G_2}\}$. The broken lines will be explained shortly. It is clear from Figure 8.1 that the probabilities and frequencies of the four states are as shown in Table 8.1.

While Table 8.1 completely describes the two-unit generation system, it is more convenient, for reasons that will become clear later in this chapter, to express the generation system model in terms of a *Capacity Outage Probability and Frequency Table* (COPAFT), as shown in Table 8.2. The second column in Table 8.2 shows the *capacity outage*, i.e., total capacity minus the available capacity; the third column shows the cumulative probability; and the fourth shows the cumulative frequency.

Observe that the cumulative probabilities in Table 8.2 can be easily obtained from the exact state probabilities in Table 8.1. Alternatively, they may be

Table 8.1 State Probabilities and Frequencies for Two-Unit System

State i	Available Capacity (MW)	Probability	Frequency
1	30	$p_1 p_2$	$p_1 p_2 (\lambda_1 + \lambda_2)$
2	20	$q_1 p_2$	$q_1 p_2 (\mu_1 + \lambda_2)$
3	10	$p_1 q_2$	$p_1 q_2 (\lambda_1 + \mu_2)$
4	0	$q_1 q_2$	$q_1 q_2 (\mu_1 + \mu_2)$

Table 8.2 COPAFT for Two-Unit System

State i	$C_O = C_i$ (MW)	$P_i = P\{C_O \geq C_i\}$	$F_i = F\{C_O \geq C_i\}$
1	0	1.0	0.0
2	10	$q_1 p_2 + p_1 q_2 + q_1 q_2 = 1 - p_1 p_2$	$p_1 q_2 \mu_2 + q_1 p_2 \mu_1$
3	20	$p_1 q_2 + q_1 q_2 = q_2$	$(p_1 q_2 + q_1 q_2)\mu_2 = q_2 \mu_2$
4	30	$q_1 q_2$	$q_1 q_2(\mu_1 + \mu_2)$

obtained by examining Figure 8.1. The state below boundary B_4 contributes to a capacity outage of 30 MW. The states below boundary B_3 contribute to a capacity outage of 20 MW and greater, and their probabilities add up to P_3. Similarly for B_2 and B_1.

The cumulative frequencies F_i can be understood as follows: F_i is the frequency of encountering states of capacity outage C_i or greater. This is equal to the frequency of transiting from states above boundary B_i to those below B_i. In the steady state, when the frequency balance condition holds, this equals the frequency of transiting from states below boundary B_i to those above B_i; often, when convenient, this frequency is calculated in determining F_i. There being no transition across B_1, $F_1 = 0$, as shown in Table 8.2.

8.2.1 The Unit Addition Algorithm

While the above method is illustrative of the nature of a COPAFT, a more practical method, called the *Unit Addition Algorithm*, is used for systems with large numbers of generating units. This is a recursive algorithm that takes an existing COPAFT and adds an additional unit to it. It does so by performing a discrete convolution of the probability and frequency distributions of the "new unit" (the unit being added) with those of the existing system. The algorithm is recursive because it begins with the first unit and continues until all units have been added. This algorithm is explained by means of the following two examples.

Example 8.1 Suppose we already have a COPAFT as in Table 8.3, and we wish to add a new unit, which is identical to the G_1 that we encountered before.

Table 8.3 COPAFT for an Existing System

State i	$C_O = C_i$ (MW)	$P_i = P\{C_O \geq C_i\}$	$F_i = F\{C_O \geq C_i\}$
1	0	1.0	0.0
2	20	q_2	$q_2 \mu_2$

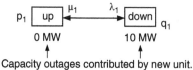

Capacity outages contributed by new unit.

Capacity outage states in existing table. These are also states assumed by modified system when new unit is up.

New capacity outage states introduced by new unit. These are states assumed by modified system when new unit is down.

Figure 8.2 Addition of 10 MW unit.

Figure 8.2 shows the existing states and the new states that are introduced into the system as a result of adding the new unit.

The addition of the 10 MW unit results in the insertion of two new outage states into the COPAFT: 10 MW and 30 MW. Consider the computation of the cumulative probability of the 10 MW outage state. The modified system, i.e., the system with the new unit included, can have an outage of 10 MW or greater in one of two ways: (a) the "original system" (without the 10 MW unit) is at outage 0 or greater and the "new unit" is down, i.e., at outage 10 MW; and, (b) the original system is at outage 20 MW and the new unit is up, i.e., at outage 0. These probability components contribute to the following:

$$P\{C_O \geq 10\} = P_1^- q_1 + P_2^- p_1 = q_1 + q_2 p_1 = (1 - p_1) + (1 - p_2)p_1 = 1 - p_1 p_2,$$
(8.1)

where the superscript "−" denotes values before the addition of the new unit, which can be found in Table 8.3. Similarly,

$$P\{C_O \geq 30\} = P_2^- q_1 = q_2 q_1,$$
(8.2)

where the second component vanishes because the original system (Table 8.3) did not have any capacity outage state greater than 20 MW. Had there been such a state, (8.2) would have been of the form

$$P\{C_O \geq 30\} = P_2^- q_1 + P_3^- p_1.$$
(8.3)

Now consider the cumulative frequencies. In general, there can be three possible transitions that result in the frequency of crossing the boundary separating

$\{C_O \geq 10\text{MW}\}$ from $\{C_O < 10\text{MW}\}$. These are marked as (A), (B) and (C) in Figure 8.2, and may be understood as follows:

(A) This transition represents the event that the new generator is at 10 MW outage, and the existing system transits from $C_O = 0$ to $C_O < 0$, so the modified system transits from $C_O = 10$ MW to $C_O < 10$ MW.
(B) This transition represents the event that the new generator is at 0 MW outage, and the existing system transits from $C_O \geq 10$ MW to $C_O < 10$ MW, so the modified system transits from $C_O \geq 10$ MW to $C_O < 10$ MW.
(C) This transition represents the event that the existing system is at 0 MW outage, and the new generator transits from $C_O = 10$ MW to $C_O = 0$, so the modified system transits from $C_O \geq 10$ MW to $C_O < 10$ MW.

These frequency components contribute to the following:

$$F\{C_O \geq 10\} = F_1^- q_1 + F_2^- p_1 + (P_1^- - P_2^-)q_1\mu_1$$
$$= 0 + (q_2\mu_2)p_1 + (1 - q_2)q_1\mu_1 = p_1 q_2 \mu_2 + q_1 p_2 \mu_1 \qquad (8.4)$$

$$F\{C_O \geq 30\} = F_2^- q_1 + 0 \cdot p_1 + (P_2^- - 0)q_1\mu_1$$
$$= (q_2\mu_2)q_1 + 0 + (q_2 - 0)q_1\mu_1 = q_1 q_2(\mu_1 + \mu_2), \qquad (8.5)$$

where, as before, the superscript "$-$" denotes values before the addition of the new unit.

Regarding this example, the following is noteworthy. The transitions labeled (A) and (B) result from changes in states of units other than the unit being added, while (C) results from a change in the state of the unit being added.

Having seen how a unit can be added to an existing system, consider the problem of using the unit addition algorithm to build the COPAFT shown in Table 8.2. It is a straightforward exercise for the reader to see that starting with zero units—see Table 8.4—and adding the 20 MW unit produces Table 8.3, and then adding the 10 MW unit, using (8.1)–(8.5), produces Table 8.2. Let us now examine what happens when the two units are added in the other sequence. This exercise illuminates aspects of the unit addition algorithm that are not clear from Example 8.1.

Example 8.2 Here we start with Table 8.4, and add the 10 MW unit, then the 20 MW unit. It is simple to add the 10 MW unit first and produce Table 8.5. Adding the 20 MW unit now has the effect shown in Figure 8.3 The probabilities

Table 8.4 COPAFT for a System with Zero Units

State i	$C_O = C_i$ (MW)	$P_i = P\{C_O \geq C_i\}$	$F_i = F\{C_O \geq C_i\}$
1	0	1.0	0.0

Table 8.5 COPAFT for Existing System with the 10 MW Unit

State i	$C_O = C_i$ (MW)	$P_i = P\{C_O \geq C_i\}$	$F_i = F\{C_O \geq C_i\}$
1	0	1.0	0.0
2	10	q_1	$q_1\mu_1$

and frequencies will be calculated as before, but for convenience the following notations will be used: The superscript "−" will be dropped from the variables denoting entries in the "old" table (Table 8.5 in this case). $P_G(X)$ and $F_G(X)$ will denote the cumulative probability and frequency for capacity outage X. Also, Table 8.5 will be assumed to be augmented by a fictitious row consisting of $C_3 = \infty$, $P_3 = 0$ and $F_3 = 0$. This will help bring out the general forms of the expressions:

$$P_G(10) = P_1 q_2 + P_2 p_2$$
$$= 1 \cdot q_2 + q_1 p_2 = 1 - p_1 p_2 \tag{8.6}$$

$$F_G(10) = F_1 q_2 + F_2 p_2 + (P_1 - P_2)q_2\mu_2$$
$$= 0 + (q_2\mu_2)p_1 + (1 - q_2)q_1\mu_1 = p_1 q_2\mu_2 + q_1 p_2\mu_1 \tag{8.7}$$

$$P_G(20) = P_1 q_2 + P_3 p_2$$
$$= 1 \cdot q_2 + 0 \cdot p_2 = q_2 \tag{8.8}$$

$$F_G(20) = F_1 q_2 + F_3 p_2 + (P_1 - P_3)q_2\mu_2$$
$$= 0 \cdot q_2 + 0 \cdot p_2 + (1 - 0)q_2\mu_2 = q_2\mu_2 \tag{8.9}$$

$$P_G(30) = P_2 q_2 + P_3 p_2$$
$$= q_1 q_2 + 0 \cdot p_2 = q_1 q_2 \tag{8.10}$$

$$F_G(30) = F_2 q_2 + F_3 p_2 + (P_2 - P_3)q_2\mu_2$$
$$= (q_1\mu_1)q_2 + 0 \cdot p_2 + (q_1 - 0)q_2\mu_2 = q_1 q_2(\mu_1 + \mu_2). \tag{8.11}$$

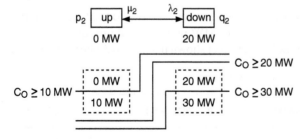

Figure 8.3 Addition of 20 MW unit.

Regarding Examples 8.1 and 8.2, the following is noteworthy. In Example 8.1, the values of $P_G(20)$ and $F_G(20)$ in the "old" table did not have to be recalculated, whereas in example 8.2, the values of $P_G(20)$ and $F_G(20)$ had to be recalculated. The reason for this is that in example 8.2, system changes resulting from transitions in the state of the unit being added (10 MW) remain within the set $\{C_O < 20 \text{ MW}\}$ and do not cross into $\{C_O \geq 20 \text{ MW}\}$, while in example 8.2, system changes resulting from transitions in the state of the unit being added (20 MW) go across the boundary of the set $\{C_O \geq 10 \text{ MW}\}$.

General Form for Two-State Units

In general, when adding a "new unit" of capacity C_N, the cumulative probability and frequency of outage state X can be determined from

$$P_G(X) = P_i p + P_j q \tag{8.12}$$

$$F_G(X) = F_i p + F_j q + (P_j - P_i) q \mu, \tag{8.13}$$

where p, q and μ are, respectively, the availability, failure probability (also known as *forced outage rate*) and repair rate of the "new unit", and P_i, P_j, F_i and F_j are determined from the "old" COPAFT as follows. i is the index of the existing capacity outage state C_i such that $C_i = X$; if such a state does not exist in the "old" COPAFT, then C_i is the smallest of the existing states that are larger than X. j is the index of the existing capacity outage state C_j such that $C_j = X - C_N$; if such a state does not exist, then C_j is the smallest of the existing states that are larger than $X - C_N$.

Equations (8.12) and (8.13) generalize the unit addition algorithm for two-state units. So a COPAFT may be constructed starting with Table 8.4 and adding one unit at a time, using (8.12) and (8.13), until all units in the system have been added. It is worth noting that the unit addition algorithm is in reality an application of the conditional probability and frequency discussed in Section 4.7. In the following section, the unit addition algorithm is extended to include three-state units.

Addition of Three-State Units

Consider adding a three-state unit, as shown in Figure 8.4, to an existing capacity outage table with outage states x_i listed in ascending order of magnitude, as

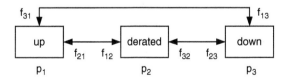

Figure 8.4 Three-state unit.

shown in Figure 8.5. In Figure 8.4, the transition frequencies are shown, rather than the transition rates as in Figures 8.2 and 8.3. These frequencies are calculated as usual, e.g., $f_{12} = p_1\lambda_{12}$. This leads to notational simplicity in the equations that follow.

Let x_i be the capacity outage in state i. The addition of a three-state unit results in three subsets of states:

$$S_1 = \{x_i\}$$
$$S_2 = \{x_i + C_D\}$$
$$S_3 = \{x_i + C_T\},$$

where

$C_T =$ capacity of unit being added;
$C_D =$ amount of capacity lost when unit being added is derated.

These subsets, arranged as three columns in Figure 8.5, have an equal number of states, and in each the capacity outages are arranged in ascending order.

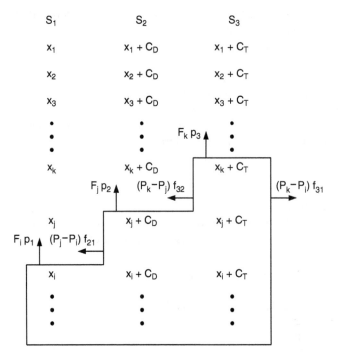

Figure 8.5 State frequency diagram for unit addition.

Assuming that a capacity equal to or greater than X is defined by states equal to and greater than i, j and k in S_1, S_2 and S_3 respectively,

$$P_G(X) = P_i p_1 + P_j p_2 + P_k p_3, \tag{8.14}$$

and

$$F_G(X) = G(X) + N(X), \tag{8.15}$$

where

$$G(X) = F_i p_1 + F_j p_2 + F_k p_3 \tag{8.16}$$

$$N(X) = (P_j - P_i)f_{21} + (P_k - P_i)f_{31} + (P_k - P_j)f_{32}, \tag{8.17}$$

and P_i, F_i = probability and frequency of capacity outage equal to or greater than x_i.

Here, $G(X)$ represents the frequency due to change in the states of the units other than the unit being added, while $N(X)$ represents the contribution of the change in the states of the unit being added.

It is instructive to see how these equations can yield the equations for adding a two-state unit. Simply ignore the third state, in both figures, and treat the derated state as the full outage state. Then the equations become

$$P_G(X) = P_i p_1 + P_j p_2 \tag{8.18}$$

$$F_G(X) = F_i p_1 + F_j p_2 + (P_j - P_i)f_{21}, \tag{8.19}$$

which are essentially the same as (8.12) and (8.13).

8.2.2 Rounding Off Generation States

Often, for large systems, the generation model (COPAFT) is constructed in increments of a convenient "step size," such as 10 MW or 100 MW. This helps keep the size of the model manageable, provided the "round-off error" is considered acceptable. An effective way of rounding off is to suitably modify the model for each unit before it is added to the existing table. The manner in which two-state and three-state units are modified is described in the following sections.

Two-State Unit
Suppose that generation system model is to be built with capacity outage states that are in multiples of 10 MW. So a two-state unit with capacity outages of 0 and 56 MW will result into a three-state unit with capacity outages of 0, 50 and 60 MW. The transition-rate diagram without roundoff is shown in Figure 8.6, where λ_{ij} represents the failure rate from state i to state j, and μ_{ij} represents the repair rate from state i to state j.

Figure 8.6 Two-state transition diagram before rounding off.

The probabilities and the frequencies of the states 1 and 2 of the unit are obtained as follows:

$$p_1 = \frac{\mu_{21}}{\lambda_{12} + \mu_{21}} \tag{8.20}$$

$$p_2 = \frac{\lambda_{12}}{\lambda_{12} + \mu_{21}} \tag{8.21}$$

$$f_{12} = \frac{\lambda_{12}\mu_{21}}{\lambda_{12} + \mu_{21}} \tag{8.22}$$

$$f_{21} = \frac{\lambda_{12}\mu_{21}}{\lambda_{12} + \mu_{21}}, \tag{8.23}$$

where

p_i = probability of the unit being in state i
f_{ij} = frequency of transition from unit state i to j.

In Figure 8.6, state 2 represents the 56 MW outage capacity. The transition-rate diagram after round off is shown in Figure 8.7. It can be seen from Figure 8.7 that the state 2 is split into two states, $2'$ and $2''$. It should be observed that there is no transition between states $2'$ and $2''$. As seen in Figure 8.7, α is the parameter that modifies the transition rates. It is the ratio of the difference between the capacity outage of state $2''$ and unit capacity outage and the increment. In this particular case,

$$\alpha = (60 - 56)/10 = 0.4$$

$$1 - \alpha = 0.6.$$

The transition rates are modified in the inverse ratio of the differences between the capacity outage state of the unit and the capacity outages in the rounded-off states. This results in the mean value of capacity outage states $2'$ and $2''$ being equal to the capacity outage of state 2. In the case of the example, since 56 is closer to 60 than to 50, the number of transitions to state 60 will be more than to state 50. But the total number of transitions to states $2'$ and $2''$ should be equivalent to the number of transitions to state 2. So the number of transitions to these modified states should be a fraction of the number of

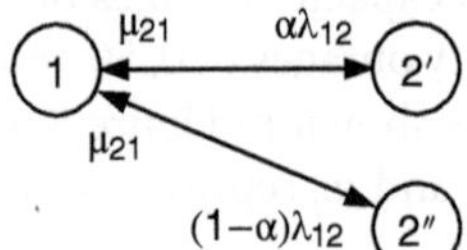

Figure 8.7 State transition diagram after rounding off.

transitions to state 2. This is obtained by multiplying the transition rate by α and $(1 - \alpha)$, which are in the inverse proportion of the differences between the exact capacity outage state and the new incremental capacity outage states. So the transition rates in Figure 8.7 are obtained as

$$\lambda_{12'} = \alpha\lambda_{12} \tag{8.24}$$

$$\lambda_{12''} = (1 - \alpha)\lambda_{12} \tag{8.25}$$

$$\mu_{2''1}, \mu_{2'1} = \mu_{21}. \tag{8.26}$$

It is clear that the repair rates are not modified, since the rate of repair remains the same in both the states. It can be verified that the sum of the transitions to states $2'$ and $2''$ in Figure 8.7 is equal to the transitions to state 2 in the Figure 8.6, i.e.,

$$\lambda_{12} = \lambda_{12'} + \lambda_{12''} = \alpha\lambda_{12} + (1 - \alpha)\lambda_{12}. \tag{8.27}$$

It can be easily shown that the mean sum of the capacity outages of the two rounded-off states is equal to the original capacity outage state, i.e.,

$$c_i = \alpha c_{i1} + (1 - \alpha)c_{i2}, \tag{8.28}$$

where

c_i = original capacity outage state
c_{i1}, c_{i2} = capacity outage states after rounding off.

Since the transition rates are modified, the probabilities and frequencies also get modified accordingly as follows:

$$p_{2'} = \alpha\lambda_{12}/(\mu_{21} + \lambda_{12}) = \alpha p_2 \tag{8.29}$$

$$p_{2''} = (1 - \alpha)\lambda_{12}/(\mu_{21} + \lambda_{12}) = (1 - \alpha)p_2 \tag{8.30}$$

$$f_{12'}, f_{2'1} = p_{2'}\mu_{21} = \alpha f_{21} \tag{8.31}$$

$$f_{12''}, f_{2''1} = p_{2''}\mu_{21} = (1 - \alpha)f_{21}. \tag{8.32}$$

After the modifications are made, the unit is added as usual using the unit addition algorithm.

Three-State Unit
In this case, a three-state unit after rounding off may result in either a four-state unit or a five-state unit depending on the capacity outages of the derated and down states. When a state of a unit is an integral multiple of the incremental step there is no need to round off that state. The various possibilities are explained below using an incremental step of 10 MW.

1. When either derated capacity outage or full capacity outage is not an integral multiple of the incremental step, then after rounding off a four-state unit will

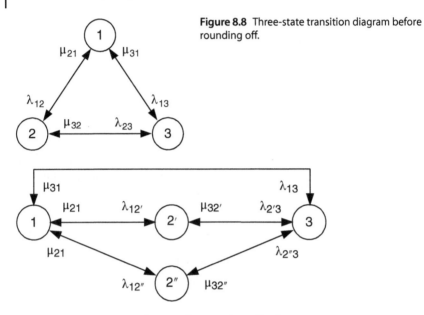

Figure 8.8 Three-state transition diagram before rounding off.

Figure 8.9 State-transition diagram after rounding off to four states.

result. For example, in the case of a unit with capacity outage states of 0, 32 and 50 MW, after rounding off there will be 0, 30, 40 and 50 MW capacity outage states, as shown in Figure 8.9.

2. When all the states are not integral multiples of the incremental step, then after rounding off a five-state unit results, as shown in Figure 8.10. An example of such a unit is one with capacity outage states of 0, 17 and 43 MW.

3. There is one more possibility that when both the derated and full capacity outage states are in the same range of the incremental step, then the resulting

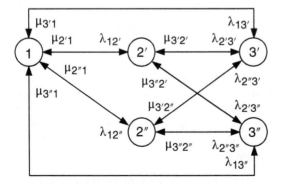

Figure 8.10 State-transition diagram after rounding off to five states.

Table 8.6 Equivalent Transition Rates for Five-State Model

Equivalent failure rates	Equivalent repair rates
$\lambda_{13'} = \alpha_2 \lambda_{13}$	$\mu_{3'1} = \mu_{31}$
$\lambda_{13''} = (1 - \alpha_2)\lambda_{13}$	$\mu_{3''1} = \mu_{31}$
$\lambda_{2'3'} = \alpha_2 \lambda_{23}$	$\mu_{3'2'} = \alpha_1 \mu_{32}$
$\lambda_{2'3''} = (1 - \alpha_2)\lambda_{23}$	$\mu_{3'2''} = (1 - \alpha_1)\mu_{32}$
$\lambda_{2''3'} = \alpha_2 \lambda_{23}$	$\mu_{3''2'} = \alpha_1 \mu_{32}$
$\lambda_{2''3''} = (1 - \alpha_2)\lambda_{23}$	$\mu_{3''2''} = (1 - \alpha_1)\mu_{32}$

unit has only three states. For example, rounding off a unit with 0, 13 and 19 MW state unit with increments of 10 MW results in a 0, 10 and 20 MW state unit. The transition rates in this case have to be modified in a slightly different way when compared to other cases.

The unit round off for a three-state unit can be described as assuming a three-state unit in which the derated and full capacity outage states are not integral multiples of the incremental step of, say, 10 MW. The state-transition diagram without rounding off is shown in Figure 8.8. The transition diagram after rounding off is shown in Figures 8.9 and 8.10. The states 2 and 3 in the Figure 8.8 are split into states $2'$, $2''$ and $3'$, $3''$, respectively. The various transition rates shown in Figures 8.9 and 8.10 are obtained as follows.

As in the case of a two-state unit, the α parameter is required. But in this case two such parameters are required. Let them be called α_1 and α_2, where α_1 is associated with the derated state, and α_2 is associated with the full capacity outage state. Both of them are calculated as described earlier. After α_1, α_2 have been obtained, the various transition rates are calculated as shown. Transition rates $\lambda_{12'}$, $\lambda_{12''}$, $\mu_{2'1}$, $\mu_{2''1}$ are obtained using the following equations, with α replaced by α_1. The other transition rates are obtained using the equations in Table 8.6.

The probabilities and frequencies of the equivalent unit can be expressed in terms of these values for the original units using the equations in Table 8.7.

Now this unit with modified probabilities and frequencies is added to the existing generation system model.

8.3 Load Model

The load model, like the generation model, consists of probability and frequency distributions of the random variable representing the total system

Table 8.7 Equivalent Probabilities and Frequencies for Five-State Model

Equivalent probabilities	Equivalent frequencies
$p_{2'} = \alpha_1 p_2$	$f_{2'1} = \alpha_1 f_{21}$
$p_{2''} = (1 - \alpha_1)p_2$	$f_{2''1} = (1 - \alpha_1)f_{21}$
$p_{3'} = \alpha_2 p_3$	$f_{3'1} = \alpha_2 f_{31}$
$p_{3''} = (1 - \alpha_2)p_3$	$f_{3''1} = (1 - \alpha_2)f_{31}$
	$f_{3'2'} = \alpha_1 \alpha_2 f_{32}$
	$f_{3'2''} = (1 - \alpha_1)\alpha_2 f_{32}$
	$f_{3''2'} = \alpha_1(1 - \alpha_2)f_{32}$
	$f_{3''2''} = (1 - \alpha_1)(1 - \alpha_2)f_{32}$

load. These distributions are expressed in discrete form, in terms of the triplet $(L, P_L(L), F_L(L))$, where for every value L_i assumed by L, $P_L(L_i) = P\{L \geq L_i\}$ and $F_L(L_i) = F\{L \geq L_i\}$. In general, construction of the load model is a straightforward process and is performed by scanning the hourly load data (observed or forecasted) of the system over the period of interest and applying (8.33) and (8.34):

$$P\{L \geq L_i\} = \frac{\text{number of hours } L \geq L_i}{\text{number of hours in interval}} \tag{8.33}$$

$$F\{L \geq L_i\} = \frac{\text{number of transitions from } \{L < L_i\} \text{ to } \{L \geq L_i\}}{\text{number of hours in interval}}. \tag{8.34}$$

The following examples illustrate the process.

Example 8.3 Consider a system in which the hourly load assumes the following values, in MW, over a 24-hour period: 54, 51, 48, 47, 47, 48, 59, 69, 76, 77, 77, 76, 76, 76, 75, 75, 79, 80, 80, 77, 73, 67, 59, 51. The load model for this system, over the 24-hour period, would be as shown in Table 8.8.

In Table 8.8, the cumulative probabilities are easy enough to understand. The cumulative frequency for any load L_i is determined by counting the number of times the load transits from $\{L < L_i\}$ to $\{L \geq L_i\}$. Load levels below 47 MW are not shown because these do not occur during the period of interest. Note that one could choose to show one or more load states between 0 and 46 MW, and it would still constitute a valid load model; of course, every one of these states would have a cumulative probability of 1.0 and a cumulative frequency of 0.0.

As with the generation model, a load model, too, might be constructed in increments of a fixed step size. The construction of such a load model is as straightforward as that of the exact state load model. This is illustrated in Example 8.4.

Table 8.8 Load Model for Example 8.3

State i	$L = L_i$ (MW)	$P_i = P\{L \geq L_i\}$	$F_i = F\{L \geq L_i\}$ (/hour)
1	47	1.0000	0.00000
2	48	22/24=0.9167	1/24=0.04167
3	51	20/24=0.8333	1/24=0.04167
4	54	18/24=0.7500	1/24=0.04167
5	59	17/24=0.7083	1/24=0.04167
6	67	15/24=0.6250	1/24=0.04167
7	69	14/24=0.5833	1/24=0.04167
8	73	13/24=0.5417	1/24=0.04167
9	75	12/24=0.5000	1/24=0.04167
10	76	10/24=0.4167	2/24=0.08333
11	77	6/24=0.2500	2/24=0.08333
12	79	3/24=0.1250	1/24=0.04167
13	80	2/24=0.0833	1/24=0.04167

Example 8.4 Consider the problem of building a load model for the same load data as given in Example 8.3, but in increments of 10 MW. This load model would be as shown in Table 8.9.

For larger data sets, a *load modeling algorithm* is used, which enables construction of the entire load model with a single scan of the hourly load data, so it is not necessary to scan the dataset once for every state in the load model. This is described below.

The Load Modeling Algorithm
The load modeling algorithm scans the hourly load data and develops a load model of the form described by (8.33) and (8.34). It is usually developed in increments of a fixed step size, Z, such that $Z = L_{i+1} - L_i$, $\forall i$. In other words, the

Table 8.9 Load Model for Example 8.4

State i	$L = L_i$ (MW)	$P_i = P\{L \geq L_i\}$	$F_i = F\{L \geq L_i\}$ (/hour)
1	40	1.0000	0.00000
2	50	20/24=0.8333	1/24=0.04167
3	60	15/24=0.6250	1/24=0.04167
4	70	13/24=0.5417	1/24=0.04167
5	80	2/24=0.0833	1/24=0.04167

load levels are $L_1 = 0, L_2 = Z, L_3 = 2Z, ..., L_i = (i-1)Z, ..., L_{n_L} = (n_L - 1)Z = L_{max}$, where n_L is the number of load levels, and L_{max} is the largest state in the load model. It is assumed that H_i is the load during the hour i and that the load model is to be built from hourly load data from hour N_1 to N_2. The load modeling algorithm consists of the following steps:

1. Determine $n_L = $ peak load$/Z + 1$.
2. Initialize $P_L(L_j) \leftarrow 0, F_L(L_j) \leftarrow 0$ for $j = 1 ... n_L$.
3. Initialize $i \leftarrow N_1$, where i is the hour under consideration.
4. Determine the contribution of the ith hour to $P_L(L_j)$:
   update $P_L(L_j) \leftarrow P_L(L_j) + 1$ for $j = 1$ to J, where $J = H_i/Z + 1$.
5. Determine the contribution to $F_L(L_j)$ due to load change from hour i to $i + 1$:
   $J_1 \leftarrow H_{i+1}/Z + 1$
   $J \leftarrow J + 1$
   (a) if $J_1 < J$, then $H_{i+1} > H_i$ and therefore there is no contribution to $F_L(L_j)$; go to 6;
   (b) if $J_1 \geq J$, then $H_{i+1} > H_i$ and there is contribution to frequency; update $F_L(L_j) \leftarrow F_L(L_j) + 1$ for $j = J$ to J_1.
6. Advance the hour, $i \leftarrow i + 1$;
   if $i \leq N_2$, go to 4.
7. $P_L(L_j) \leftarrow P_L(L_j)/N_H$
   $F_L(L_j) \leftarrow F_L(L_j)/N_H$,
   where $N_H = N_2 - N_1 + 1$.

This algorithm results in the probability vector $P_L(L)$ and the frequency vector $F_L(L)$, the unit of the latter being h^{-1} (i.e., per hour). The algorithm is efficient and suitable for computer implementation and builds the entire load model from a single scan of the hourly load data.

8.4 Generation Reserve Model

This model consists of the probability and frequency distributions of the random variable M, representing the generation reserve margin. In terms of the generation and load models, described in the previous sections, M can be expressed as:

$$M = C_C - C_O - L, \tag{8.35}$$

where C_C is the total generation capacity available for commitment, i.e., the installed capacity minus the capacity on planned outage, C_O is the capacity on forced outage and L is the system load.

Clearly, M represents the amount of generation capacity, in excess of the load, that is available for commitment. The probability distribution function of M can be expressed as:

$$P\{M \leq M_i\} = \sum_{j=1}^{n_G} P\{C_O = C_j\} P\{L \geq C_C - C_j - M_i\}, \tag{8.36}$$

where n_G is the number of states in the generation model (the COPAFT). This can be understood as follows. When $C_O = C_j$, the jth capacity outage level, then the "capacity in" is $C_C - C_j$, and $C_C - C_j - M_i$ equals the load, because of (8.35). So when the load L equals or exceeds $C_C - C_j - M_i$, the margin M is less than or equal to M_i. The first term on the right in (8.36) can be obtained from the COPAFT: $P\{C_O = C_j\} = P_G(C_j) - P_G(C_{j+1})$, the difference between the cumulative probabilities of the jth and $(j + 1)$th capacity outage states yielding the exact state probability of C_j. The second term on the right comes from the load model: $P\{L \geq C_C - C_j - M_i\} = P_L(C_C - C_j - M_i)$.

The product on the right in (8.36) is calculated for each capacity outage level, and the sum of these products yields the cumulative probability for margin M_i. This probability is calculated for every possible value of M_i, thus completing the discrete convolution between the random variables $C_C - C_O$ and L.

The frequency distribution function of M is given by:

$$F\{M \leq M_i\} = F^g\{M \leq M_i\} + F^\ell\{M \leq M_i\}, \tag{8.37}$$

where

$$F^g\{M \leq M_i\} = \sum_{j=1}^{n_G} F\{C_O = C_j\} P\{L \geq C_C - C_j - M_i\} \tag{8.38}$$

$$F^\ell\{M \leq M_i\} = \sum_{j=1}^{n_G} P\{C_O = C_j\} F\{L \geq C_C - C_j - M_i\}. \tag{8.39}$$

In (8.37), the first term on the right is the frequency of encountering system states with margin M_i or less as a result of generation system transitions only and is determined using (8.38). In (8.38), each term on the right equals the frequency of encountering generation states of capacity equal to $C_C - C_j$ multiplied by the probability of the load being equal to or exceeding $C_C - C_j - M_i$. The first term is obtained from the COPAFT: $F\{C_O = C_j\} = F_G(C_j) - F_G(C_{j+1})$ and the second term from the load model.

In (8.37), the second term on the right is the frequency of encountering system states with margin M_i or less as a result of load transitions only and is determined using (8.39). In (8.39), each term on the right equals the probability of encountering generation states of capacity equal to $C_C - C_j$ multiplied by the frequency of the load being equal to or exceeding $C_C - C_j - M_i$. As before, the first term is obtained from the COPAFT and the second term from the load model.

In summary, construction of the generation reserve model from the generation and load models consists of the following steps:

1. The largest positive margin possible is the largest value of $C_C - C_O$, and the smallest negative margin (magnitudinally largest) is $-L_{max}$, where L_{max} is the largest state in the load model. The margin M attains values over the range $C_C - C_O$ to $-L_{max}$. Let there be n_M such states.
2. For each value M_i of the margin, $i = 1, 2, \ldots, n_M$, the cumulative probability $P_M(M_i) = P\{M \leq M_i\}$ and the cumulative frequency $F_M(M_i) = F\{M \leq M_i\}$ are determined using (8.40) and (8.41), which are derived from (8.36)–(8.39).

$$P\left(M_i\right) = \sum_{j=1}^{n_G} \left[P_G(C_j) - P_G(C_{j+1})\right] P_L\left(C_C - C_j - M_i\right) \tag{8.40}$$

$$F\left(M_i\right) = \sum_{j=1}^{n_G} \left\{\left[F_G(C_j) - F_G(C_{j+1})\right] P_L\left(C_C - C_j - M_i\right)\right.$$
$$\left. + \left[P_G(C_j) - P_G(C_{j+1})\right] F_L\left(C_C - C_j - M_i\right)\right\} \tag{8.41}$$

The complete generation reserve model is defined by the triplet $(M, P(M), F(M))$. However, if this is to be used only for the determination of generation adequacy indices, then it is sufficient to build the model over the range 0 to $-L_{max}$. Moreover, in practice, the COPAFT may be truncated at C_{jmax} if $P\{C_O \geq C_j\}$ becomes insignificantly small for $C_j > C_{jmax}$. In this case the smallest value of M that needs to be considered is $M_{min} = C_C - C_{jmax} - L_{max}$, instead of $M_{min} = -L_{max}$. The indices are determined as shown in the next section.

8.5 Determination of Reliability Indices

From the generation reserve model, the system reliability indices are computed as follows:

$$\text{LOLP} = P(-M_0) \tag{8.42}$$
$$\text{HLOLE} = \text{LOLP} \times D \tag{8.43}$$
$$\text{LOLF} = F(-M_0) \tag{8.44}$$

$$\text{EPNS} = (\Delta M) \left[\sum_{M=0}^{-L} P(M) - \frac{1}{2}\{P(0) + P(-L)\}\right] \tag{8.45}$$

$$\text{EUE} = \text{EPNS} \times D, \tag{8.46}$$

where

$$
\begin{aligned}
\text{LOLP} &= \text{loss of load probability} \\
\text{HLOLE} &= \text{hourly loss of load expectation} \\
\text{LOLP} &= \text{loss of load frequency} \\
\text{EPNS} &= \text{expected power not served} \\
\text{EUE} &= \text{expected unserved energy} \\
M_0 &= \text{magnitudinally smallest margin} \\
\Delta M &= \text{increment at which } P(M) \text{ is computed} \\
D &= \text{length of time interval in hours} \\
-L &= M_{min} \\
&= \text{smallest negative margin considered}
\end{aligned}
$$

The value of M_0, the smallest margin, is usually the same as the increment ΔM; however, some choose $M_0 = 0$, regarding the state where the reserve margin is zero as potentially a loss of load state. The value of $-L$ or M_{min}, the smallest negative margin considered, is given by $M_{min} = C_C - C_{jmax} - L_{max}$, as described in Section 8.4.

Equations (8.42) and (8.44) are understood easily enough, realizing that loss of load occurs as soon as the margin becomes negative. In (8.43), HLOLE is an expectation index, showing the expected number of hours during which loss of load occurs over the period of interest. In (8.45) and (8.46), the EPNS and EUE are severity indices, indicating the expected amounts of power and energy curtailed over the period of interest. EPNS and EUE are typically expressed in MW/year and MWh/year, respectively.

The determination of EPNS using (8.45) can be understood as follows. Consider Figure 8.11, which shows the probability distribution function of the generation reserve margin M over the domain where $M \leq 0$. The EPNS corresponds to the area under the probability distribution curve over this domain and approximately equals the sum of the areas of the shaded rectangles. This sum can be calculated as follows:

$$
\begin{aligned}
\text{EPNS} &= \frac{\Delta M}{2}\{P(0) - P(-\Delta M)\} + \frac{3\Delta M}{2}\{P(-\Delta M) - P(-2\Delta M)\} \\
&\quad + \frac{5\Delta M}{2}\{P(-2\Delta M) - P(-3\Delta M)\} + \cdots \\
&\quad + \left(L - \frac{\Delta M}{2}\right)\{P(-L + \Delta M) - P(-L)\} + L\{P(-L)\} \\
&= \frac{\Delta M}{2}P(0) + \left(\frac{3\Delta M}{2} - \frac{\Delta M}{2}\right)P(-\Delta M) + \left(\frac{5\Delta M}{2} - \frac{3\Delta M}{2}\right)P(-2\Delta M) \\
&\quad + \cdots + \left\{\left(L - \frac{3\Delta M}{2}\right) - \left(L - \frac{\Delta M}{2}\right)\right\}P(-L + \Delta M) + \frac{\Delta M}{2}P(-L) \\
&= \frac{\Delta M}{2}P(0) + (\Delta M)[P(-\Delta M) + P(-2\Delta M) + \cdots + P(-L + \Delta M)] \\
&\quad + \frac{\Delta M}{2}P(-L)
\end{aligned}
$$

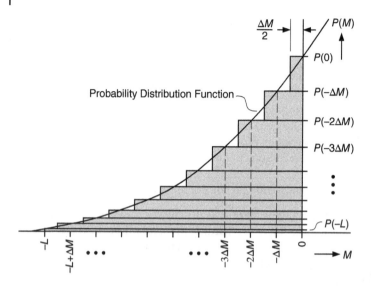

Figure 8.11 Probability distribution function of generation reserve margin.

$$= (\Delta M)\,[P(0) + P(-\Delta M) + P(-2\Delta M) + \cdots + P(-L + \Delta M) + P(-L)]$$
$$-\frac{\Delta M}{2}\,\{P(0) + P(-L)\}$$

$$= (\Delta M)\left[\sum_{M=0}^{-L} P(M) - \frac{1}{2}\,\{P(0) + P(-L)\}\right].$$

This is identical to (8.45).

8.6 Conclusion

In this chapter we dealt with the construction of discrete state generator and load models and with the process of discrete convolution. Discrete state models use the best available data for individual components. It is worthwhile to mention that there have been efforts to construct other models for generators. Some of these have used continuous distribution approximations, such as the Gram-Charlier method [29], Edgeworth method [30], mixture of normals [31], Leguerre polynomials [32], Legendre series [33] and multi parameter gamma distribution [34]. These developments were primarily motivated by the need to improve computational speed, but subsequent improvements in computing hardware have reduced the need to resort to these approximate methods, and discrete distributions are widely used in the present day.

It is customary to consider the forced outages and repair of different generators as independent events, except in common-mode or common cause events.

These events are discussed in Chapter 9, which deals with reliability models for interconnected systems. Even in interconnected systems, discrete generation models as described in this chapter are used. Interconnected systems also use discrete load models, but it is often necessary to consider the temporal correlations between nodal loads, and this is also covered in Chapter 9.

Exercises

8.1 Manually perform the following simple exercise in generation adequacy evaluation using discrete convolution.

(a) Build a capacity outage table with probability and frequency distributions for a system consisting of the following three generators:

Capacity (MW)	λ (/y)	μ (/y)
30	5	20
20	15	35
50	3	27

(b) Build a load model for the first 24 hours in the IEEE Reliability Test System (IEEE-RTS) [35] load curve, using 85 MW for the annual peak load and 10 MW intervals. (The loads, over the 24-hour period, are: 45.7, 42.9, 40.9, 40.2, 40.2, 40.9, 50.4, 58.6, 64.7, 65.4, 65.4, 64.7, 64.7, 64.7, 63.4, 64.1, 67.5, 68.1, 68.1, 65.4, 62.0, 56.6, 49.7, 42.9 MW.)

(c) Build the generation reserve model for this system using 10 MW intervals.

(d) Calculate the LOLP, LOLF (in failures/d) and EUE (in MWh/d) for this case.

8.2 Write a computer program to perform the following for the IEEE Reliability Test System (IEEE-RTS) [35].

(a) Construct the generation system model in the form of a capacity outage table with cumulative probabilities and frequencies. Ignore states with cumulative probability less than 10^{-7}.

(b) Construct the load model with cumulative probabilities and frequencies.

(c) Construct the generation reserve model by performing a discrete convolution of the generation system model and the load model. Ignore states with cumulative probability less than 10^{-7}.

(d) From the generation reserve model, determine the HLOLE, the LOLF and the EUE for the IEEE-RTS.

9

Reliability Analysis of Multinode Power Systems

9.1 Introduction

This chapter introduces reliability measures and component models used in multinode power systems. Many of the measures and models are the same as those described in Chapter 8; however, there are some additional considerations to be taken into account in the analysis of bulk power systems, such as load correlations, transmission system modeling and operating policies, and these will be described in this chapter. The next two chapters will describe some of the methods used in reliability evaluation of multinode systems.

9.2 Scope and Modeling of Multinode Systems

This text is concerned with the evaluation of bulk power system reliability indices. In reliability evaluation of the bulk power system (generation and transmission), two kinds of analyses are used depending on the extent and the purpose of the study: composite system analysis is used to perform a detailed study of a system or a part thereof, while a multi-area study usually encompasses a much larger system of interconnected utilities or control areas. Composite power system reliability evaluation is concerned with the total problem of assessing the ability of the generation and transmission system to supply adequate and suitable electrical energy to major system load points. Whereas in multi-area reliability evaluation the transmission constraints are only indirectly considered, in composite power system reliability internal transmission limitations are directly modeled. Composite system reliability studies can be useful for better representation of generation effects in transmission system reliability analysis, or to optimize relative investments in generation and transmission systems, or for including dispersed generation.

Multi-area and composite system reliability studies are both multinode and similar in many ways. The major difference is in the transmission network modeling; because of more detailed network model, the composite system reliability

Electric Power Grid Reliability Evaluation: Models and Methods, First Edition. Chanan Singh, Panida Jirutitijaroen, and Joydeep Mitra.
© 2019 by The Institute of Electrical and Electronic Engineers, Inc. Published 2019 by John Wiley & Sons, Inc.

model has many more nodes than the multi-area model. Also, the type of network flow representations used in these studies are sometimes different. Network flow (transportation) model and DC flow methods are considered adequate for multi-area reliability evaluation, but DC flow or AC flow methods are considered more appropriate for composite system reliability evaluation. These models are discussed later in this chapter.

The methods used in multinode power system reliability evaluation can be either deterministic or probabilistic. The main principle in the deterministic methods is to maintain adequate service under most likely outages but to accept some degradation of performance under low-probability outages involving multiple generation and transmission facilities. In probabilistic methods the factors affecting reliability are modeled more comprehensively and may be analytical, Monte Carlo simulation-based or hybrid.

We will first discuss the component models and then the various methods of evaluation of reliability.

9.3 System Modeling

The reliability of an interconnected system is affected not only by the capacities and reliabilities of the individual components but also by such issues as operating policies, firm contracts and government legislation. Techniques for reliability evaluation of multinode systems have attempted to model and incorporate some of these issues in addition to modeling and integrating the system components and topology.

Consider an interconnected system consisting of four areas connected in a loop and with a tie-line across areas A and C. The term *area* is used in an arbitrary manner to represent either a utility or part of a utility. Depending upon the needs of the study, the system can be represented as multi-area comprising four nodes and five arcs or as a composite system with all the nodes, as shown in Figure 9.1. Each node represents an area or a bus, with generation and load models as described below; each arc represents an equivalent tie between areas, or a transmission link—a line or a transformer. The modeling considerations involving such representations are briefly discussed below.

The resulting network consists of N_n nodes and N_a arcs connecting these nodes. In a multi-area representation, each node represents an area, and the arcs between the nodes represent the tie-lines. Each area is represented by a single node to which the generation and load are connected. This does not mean that the intra-area constraints are ignored. The tie-lines between areas represent equivalent ties, and the intra-area bottlenecks are, to a certain degree, reflected into the tie-line capacities. Power flow studies are performed to determine the capacities of the tie lines, and these capacities are also governed by the intra-area constraints. In composite system studies, a similar

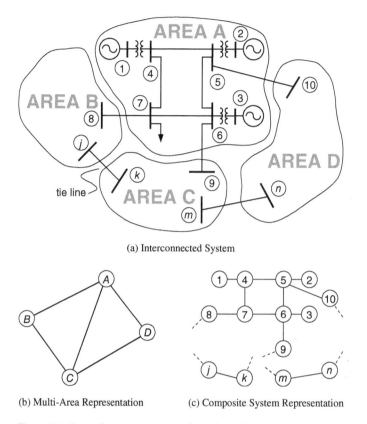

(a) Interconnected System

(b) Multi-Area Representation   (c) Composite System Representation

Figure 9.1 Network representations of a multi-node system.

network representation is used, with the nodes representing the buses, and the arcs representing the transmission lines.

9.3.1 Component Capacity States

At any given time a component, such as a generator or a transmission line, can exist in one of several capacity states: it may be fully available, it may be on outage or it may be in one of several intermediate derated states. An *outage* refers to a state when an equipment is taken out of service. Two kinds of outages are considered: planned and forced. Equipment is on planned outage when it is taken out for scheduled maintenance or preemptive repair. Forced outage occurs in the event of a random failure. Equipment may be in a derated state due to a part of it being on outage, or due to climatic conditions.

When information about outages or deratings is available in advance, such information is used by dividing the study period into intervals over which the outage or derated states persist. During each such interval, therefore, changes in component states are assumed to result only from random failures. The probabilities of these failures must, of course, be known in order to enable analysis.

9.3.2 Nodal Generation

Conventional generators are treated as dispatchable. Dispatchable resources are those that produce power upon demand, in response to control signals issued by the system operator, limited only by its capacity. On the other hand, most renewable energy resources, such as wind and solar, are considered variable or nondispatchable, because their availability is usually subject to the whims of nature.

Dispatchable generators are often modeled as two or three state devices, as described in Chapter 8. Based on the capacities and probabilities of these states, a probability distribution is constructed for the available generation in that area (or bus). This distribution may be discrete, or it may be approximated by a continuous distribution. A discrete distribution is usually obtained in the form of a *capacity outage table*, constructed using the *unit addition algorithm* as described in Chapter 8. Continuous distribution approximations have used Gram-Charlier expansion [29], Laguerre polynomials [32], Legendre series [33] and multiparameter gamma distributions [34]. Distributions of different generators, as well as those of different nodal generations, are treated as independent.

Nondispatchable generation is mostly found as cohorts of variable energy resources (VER), and their models are discussed in Chapter 13. Usually when a VER cohort, such as a wind farm or solar park, is located at a node, a multistate model is developed for the entire cohort and connected to the node. If dispatchable generation is also connected to the same node, its distribution is convolved with the multistate distribution of the VER cohort to form an equivalent capacity outage table representing the generation available at the node.

9.3.3 Nodal Loads

The nodal load represents the total load connected to that node. The loads are correlated with the time of day. There is an autocorrelation within the time series of the loads at a node, and there are correlations between nodal loads. The loads should be treated as neither independent nor perfectly correlated with coincident peaks. The diversity in the peaks and troughs is in fact a reason for interconnection of the areas.

The multinode hourly load for N_n nodes may be denoted as

$$\mathbf{L^i} = \left(L_1^i, L_2^i, \ldots, L_{N_n}^i \right) = \text{load for hour } i,$$

where

L_j^i = load in area j for hour i,

One possible way to model these correlated loads is to use a joint probability distribution. The joint probability distribution $g(L_1, L_2, \ldots, L_n)$ is such that

$$P\{l_1 < L_1 \le l_1 + \Delta l_1, l_2 < L_2 \le l_2 + \Delta l_2, \ldots l_n < L_n \le l_n + \Delta l_{N_n}\} = g(l_1, l_2, \ldots l_{N_n}) \Delta l_1 \Delta l_2 \cdots \Delta l_{N_n},$$

The derivation and utilization of such a joint distribution is, however, extremely difficult. It has been shown [36] that a more convenient approach is to use a discrete joint distribution. The hourly load vectors are grouped into a suitable number (usually between 10 and 20) of clusters [37, 38], using nearest centroid sorting. Each cluster centroid is used as an equivalent load vector and is associated with a probability based on the number of hourly load vectors that are grouped into that cluster. Since the hourly load vectors are clustered rather than the individual area load points, the cluster model also reflects the correlation between area loads. This multinode cluster load model may be represented by discrete vectors

$$\mathbf{d}^i = \left(d_1^i, d_2^i, \ldots, d_{N_n}^i \right),$$

where

d_j^i = load at node j for state i,

The probability mass function is given by

$$P\{\mathbf{D} = \mathbf{d}^i\} = \text{Probability of load state } i.$$

It has been shown [36, 38, 39] that for multinode reliability studies this proves to be a reasonable representation.

9.3.4 Transmission Line and Tie-Line Models

In multi-area studies, equivalent tie-lines are used. An equivalent tie has several capacity levels, each level associated with a probability. Such a set of capacity states with a probability mass function provides adequate representation of inter-area ties for most multi-area studies. However, depending on which power flow model is deemed appropriate, additional information concerning the line impedances corresponding to the capacity states may be required. It is often difficult to compute equivalent impedances for tie-line representations [40–42]. In most cases, it is sufficient to capture the relative values of the impedances, so that the flows will be distributed as they would be if the original transmission lines were used.

In composite system studies, the capacities and impedances of transmission lines are available, and this data, along with the probability mass functions of the discrete capacity states, comprises the transmission line model. Distributions of tie-line and transmission line states are treated as independent.

Transmission lines are assumed to be either in the up state or failed state. The failure and repair rates are further assumed to be dependent on weather. The state-transition diagram of a transmission line is shown in Figure 9.2 below, where

λ =   failure rate in normal weather;
λ' =   failure rate in adverse weather;
μ =   repair rate in normal weather;
μ' =   repair rate in adverse weather;
N, S =   mean durations of normal and adverse weather.

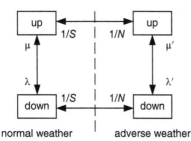

Figure 9.2 Weather-dependent model of transmission line.

Although weather can exist in several states, it is typically assumed to exist either in the normal or adverse state. An important issue concerning weather is its extent of coverage area. When the system is spread over a large area, at any given time different regions may have different states of weather. An exact treatment of this effect is difficult, and some simplifications are needed. One approach divides the whole area into regions. The weather in each region is characterized by mean duration of normal and adverse states. The weather changes in different regions are assumed to be independent. Every line is assigned to a particular region, which really means that this line is predominantly effected affected by the weather in this region. It should be noted that for parallel transmission lines just using a failure rate that is average of normal and adverse weather rates can lead to error.

9.3.5 Transformers and Buses

Like the transmission lines, transformers and buses are treated as two-state devices, but the failure rate and repair rates are assumed to be independent of the weather.

9.3.6 Circuit Breakers

A circuit breaker can have several failure modes, as described below:

1. Ground fault: This refers to the fault in the circuit breaker itself. For this type of fault, the circuit breaker is treated in the same fashion as a transformer or a bus.
2. Failure to open: The objective of a circuit breaker is to isolate the faulted component. Because of latent or hidden faults in the breaker or the associated protection system, the breaker may not open when needed. This may result in healthy components being isolated due to the operation of secondary zone protection. This failure mode is typically characterized by a probability q_{CB} in reliability modeling. This means that when this breaker receives a command to open, there is a probability q_{CB} that it may not respond. This could be either due to a problem in the breaker or in the associated protection system.
3. Undesired tripping: It is also possible that a breaker may open without a command or fault. This can be characterized as a rate and its effect will be an open line.

9.3.7 Common Mode Failures

A common mode outage is an event where multiple outages occur because of one external cause. An example of a common mode outage is the failure of two parallel circuits resulting from an ice storm; another example is the failure of a transmission tower supporting two circuits. Although several common mode outage models have been proposed, some simple common mode outage models for two components are shown in Figure 9.3. In this figure, T_i indicates that line i is up, while $\overline{T_i}$ indicates that line i is down; λ_i and μ_i are the failure and repair rates of components i and λ_{cm}, and μ_{cm} are the common mode failure and repair rates. The difference between the two models is as follows. The four-state model is applicable where the same cause results in the simultaneous outage of both components, but the components are repaired independently, so that

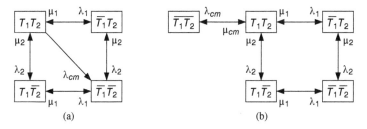

Figure 9.3 Common-mode failure models: (a) four-state; (b) five-state.

there is no "common mode repair"; in the five-state model, the common-mode failure state is distinct from the overlapping failure state, and the repair and subsequent restoration of both components is considered a single event. The first example of the ice-storm damage might be represented by either model, depending on the method of repair and restoration, while the second example of a failing tower may be more appropriately represented by the five-state model.

9.3.8 The State Space

For any given load scenario, the available area generations and the tie-line capacities will determine whether or not the area loads will be satisfied. We can thus define a state space as the set of all possible combinations of generation levels and transmission capacities. In general, therefore, a system with N_n nodes and N_a arcs will have a discrete state space of dimension $(N_n + N_a)$, with each axis consisting of the area generation or transmission capacity levels, zero levels included.

In the case of a composite system, the state space axes correspond to bus generation levels and transmission line capacity levels.

9.4 Power Flow Models and Operating Policies

Since a multi-area reliability study must consider the transmission network, the power flow model is an important issue. The following power flow models are in use:

1. The *capacity flow* model, alternatively known as the *Transportation* model, uses only the conservation equations and the tie-line capacity constraints.
2. The *DC load flow* model uses the linearized *P-δ* equations, recognizing only the tie-line susceptances, and the tie capacity constraints.
3. The *AC load flow* model uses either the complete set of nonlinear AC load flow equations, or linearized *P-δ*, *Q-V* equations, and the tie capacity constraints.

These models are briefly described below.

9.4.1 Capacity Flow Model

This model is also called a transportation model since it represents magnitude of real power that a component can transport or supply. The model is based only on Kirchhoff's first law and is given in the following:

$$AF + G = D, \tag{9.1}$$

where

A = Node-branch incidence matrix;
F = Circuit flow vector;
G = Generation vector;
G_{max} = Maximum available generation vector;
D = Demand vector.

The constraints are shown in the following:

$$G \leq G_{max} \tag{9.2}$$

$$F \leq F_f \tag{9.3}$$

$$-F \leq F_r. \tag{9.4}$$

Constraint (9.2) represents the maximum available capacity that a generating unit can supply. Constraints (9.3) and (9.4) represent maximum forward and backward flow of a transmission line, where subscripts "f" and "r" refer to the forward and reverse maximum possible flows. This model is generally considered adequate for multi-area reliability analysis, but for composite system reliability it is not considered accurate enough, as it does not include the constraints imposed on the distribution of power by the line admittances. In applications, the model is usually solved using a labeling algorithm [18]. This model has been used in some commercial programs for multi-area reliability evaluation.

9.4.2 DC Flow Model

This model considers only real power flow calculation, which is a linearized version of power flow equations. This model is given by the following equation:

$$B\delta + G = D, \tag{9.5}$$

where

b_{ij} = Negative of susceptance between nodes i and j, $i \neq j$;
b_{ii} = Summation of susceptances connected to node i;
δ = Node voltage angle vector.

The constraints in this model are the same as given in (9.2)–(9.4). As seen in (9.5), the major difference between linear network flow model and DC load flow model is that the flow calculation includes the consideration of the susceptance matrix. The line flow from node i to node j is given by (9.6):

$$F_{ij} = (\delta_j - \delta_i)b_{ij}. \tag{9.6}$$

9.4.3 AC Flow Model

The AC flow model handles both active and reactive power aspects. In addition to the bounds on generator active power generation, there are bounds to reactive power and constraints on line flows and bus voltages.

9.4.4 Firm Interchanges

A firm power interchange contract between two areas i and j obligates the exporting area i to deliver d_{ij} MW of firm power to the importing area j. Firm interchanges have a temporal relationship with the load. Since these interchanges use up a certain amount of tie-line capacity, the tie capacity becomes a function of time. In normal multi-area simulations, tie capacities are assumed to vary randomly because of component failures, and therefore this poses an additional concern. If the firm interchange lasts the entire day but varies on weekly or monthly basis, then the problem is more easily dealt with. If, however, the interchange lasts only a part of the day, then the problem becomes more complicated, and sequential Monte Carlo simulation may be a more appropriate way to deal with it. For mid-term to long-term planning, the assumption of constant firm interchange provides reasonable approximation.

9.4.5 Emergency Assistance

In the event of loss of load, two emergency action policies are recognized:

1. The *load loss sharing* (LLS) policy, where areas share unserved demand to the extent possible.
2. The *no load loss sharing* (NLLS) policy, where each area attempts to meet its own demand first. If there is excess capacity, it is supplied to the neighboring areas according to a specified priority list.

In most multi-area agreements, the NLLS policy is exercised. In composite system studies, these constraints do not enter into the modeling considerations. Generation is despatched in accordance with whatever feasible flow minimizes the net system curtailment. In a mathematical sense, this is identical to the LLS model.

9.4.6 The Remedial Action Model

All of the above models and considerations concerning power flow in the network and the impact thereon of system operating policies are captured in what is known as the remedial action model, also known as the redispatch model. This model constitutes an optimization problem that seeks to find a feasible dispatch that meets the system load while satisfying all security constraints. This optimization problem assumes the following form:

Min/Max *Objective function.*
Subject to
Power balance constraints (equality),
Security constraints (inequality).

A feasible redispatch/remedial action scenario is sought by minimizing the total system curtailment, formulating the problem as [43, 44]:

$$P_C = \text{Min} \left(\sum_{i=1}^{N_n} w_i P_{Ci} \right), \tag{9.7}$$

where P_{Ci} and w_i are the power curtailed at node i and the weight thereof, based on its priority, and N_n is the number of nodes. Variations exist, ranging from using unity weights to splitting the load at each node and assigning them different weights. Another variant of the above formulation is of the form [45]:

$$P_D = \text{Max} \left(\sum_{i=1}^{N_n} v_i P_{Di} \right), \tag{9.8}$$

where P_{Di} and v_i are the connected load at node i and the fraction of this load that the dispatch can satisfy.

For a given system state, if the total curtailment P_C obtained from the solution is zero (or, equivalently, P_D equals the total system load), then a feasible security-constrained dispatch has been found and there is no system loss of load. Otherwise, the prevailing state is considered a failure or loss of load state.

The power balance constraints are formulated using the capacity flow, DC flow or AC flow model. The corresponding models are also reflected in the line flow constraints that comprise part of the security constraints. The remaining inequality constraints are the generating capacity constraints. In an AC flow model, the equality constraints include reactive power balance, and the inequality constraints include generator reactive power limits and nodal voltage magnitude limits; none of these constraints are considered in the capacity flow model or DC flow model.

Consequently, when the capacity flow model or DC flow model is used, the optimization problem becomes linear and can be solved using robust linear programming tools [43, 44]. When using the AC flow model, the constraints become nonlinear, and nonlinear programming tools must be employed.

The system operating considerations, i.e., firm contracts and emergency assistance policy, are incorporated into the constraint set.

10

Reliability Evaluation of Multi-Area Power Systems

10.1 Introduction

This chapter provides an overview of the different methods that are used in the reliability analysis of multi-area power systems. One of these methods, that of state space decomposition, is explained in some detail in order that the reader may get a better understanding of the problem. These methods draw upon the fundamental concepts of system reliability modeling and analysis that were covered in Part I, and use or build upon the component models described in Chapters 8 and 9.

10.2 Overview of Methods for Multi-Area System Studies

In this section we provide brief descriptions of the various methods used for reliability analysis of multi-area power systems. In the next section we shall describe one of these methods—the method of state space decomposition—to provide a deeper understanding of the various aspects of multi-area reliability evaluation.

10.2.1 Contingency Enumeration/Ranking

Contingency enumeration [46, 47] consists of listing all contingencies that result in system failure and aggregating their contribution to the reliability indices. In the context of multi-area reliability evaluation, *system failure* implies that the system is not able to supply all the load at one or more areas. This is a brute force method and is highly expensive in terms of computational effort. Two approaches have been used to alleviate the computing burden:

1. Ranking failure states by severity: The failure states are ranked by performing a base case load flow and then ranking them by a performance index [48]. More information on this method is provided in Chapter 11.

Electric Power Grid Reliability Evaluation: Models and Methods, First Edition. Chanan Singh, Panida Jirutitijaroen, and Joydeep Mitra.

2. Discarding high-order (low-probability) contingencies from the state space [28].

10.2.2   Equivalent Assistance Approach

This method [4] is easily understood for a two-area system, where the probability distribution of the equivalent assistance available from one area is convolved with that of the generation reserve of the other, and the reliability indices are obtained from the resulting distribution. The extension to the radially connected multi-area case is obvious [49]. This technique, however, does not easily extend to looped configurations; [50] determines capacity assistance for arbitrary network configurations but uses state space decomposition as a tool.

10.2.3   Stochastic/Probabilistic Load Flow

This approach begins by estimating the probability distributions of voltage magnitudes and line flows in a network based on the known distributions of generation and load. The distributions of voltages and flows are then used to determine probabilities of constraint violations and thereby compute reliability indices. The difference between stochastic [51] and probabilistic [52] approaches is that while the former uses such parameters as expectations and variances to describe the distributions, the latter uses the discrete distributions directly.

10.2.4   State Space Decomposition

State space decomposition [39, 43, 47, 53–56] is an analytical method that recursively decomposes the system state space into disjoint sets of acceptable, unclassified and loss of load states. The reliability indices are then determined from the loss of load sets. State space decomposition can prove to be a very efficient method due to the fact that it deals with sets of states rather than individual states. However, since this method deals with sets of states, coherence is a required condition, and so most methods use only the capacity flow model. In some instances [43, 53, 57] a DC flow model has been used, either with some limitations or as part of a hybrid approach. A description of the decomposition method appears later in this chapter.

10.2.5   Monte Carlo Simulation

The Monte Carlo method mimics the history of the system operation by using the probability distributions of the component states. Statistics are then collected and indices estimated using statistical inference. Chapter 6 provides a description of the Monte Carlo method, and Chapter 11 describes its application to power system reliability.

10.2.6   Hybrid Methods

Hybrid methods consist of combinations of two or more of the methods outlined above. These methods are so selected as to compensate for each other's weaknesses. The most favored combination has been that of decomposition and simulation. Decomposition is extremely effective during the initial recursions because of its ability to classify large sets. However, after several layers of decomposition, the efficiency diminishes, because the probabilities of the sets classified become progressively smaller, while the numbers of such low-probability sets become progressively larger. For this reason it has been found expedient to discontinue decomposition at a suitable stage and switch to Monte Carlo Simulation. In some hybrid methods [43, 55, 56] part of the indices are determined from decomposition, the rest from simulation; in others [9, 57] decomposition removes only acceptable states, leaving all the failure states aside, so that the indices are entirely determined from simulation.

10.3   The Method of State Space Decomposition

In the remainder of this chapter, the method of state space decomposition is described in detail. Examples and graphical interpretations are provided to supplement the verbal descriptions.

10.3.1   State Space Decomposition

Consider the multi-area space as described in Section 9.3.8. This state space has been defined for a single load scenario. The method of state space decomposition, per se, applies to only a single load scenario. The manner in which this method is extended to accommodate temporal load variations, such as in the inclusion of an annual load curve, will be dealt with in the section describing the method of simultaneous decomposition. The current section deals only with the method of decomposition as applied to a single load scenario.

As stated before, the state space can be defined as the set of all possible combinations of generation levels and tie-line capacities. For a given load scenario, the state space can be recursively decomposed [55] into A-sets (acceptable sets, comprising states which do not result in loss of load), L-sets (loss of load sets) and U-sets (unclassified sets). In the first sweep, the entire state space is regarded as an U-set, and, based on its highest and lowest capacity states, is decomposed into an A-set, L-sets and U-sets. In subsequent sweeps, each U-set is further decomposed into an A-set, L-sets and more U-sets. This is continued until no U-set remains. The total probability of all the L-sets is the system Loss of Load Probability (LOLP). Every L-set is also decomposed into B-sets (comprising identical area loss of load states) and W-sets (of unclassified loss of load states), and every W-set is further decomposed into B-sets and more W-sets, until at the end only A-sets and B-sets remain. The B-sets yield the area LOLPs and area EUEs (Expected Unserved Energy). The system EUE equals the sum of the area EUEs.

The process of decomposition ensures that sets generated from every decomposition are disjoint, and each set is represented by two *boundary states*. The identification of boundary states is accomplished by determining *partition vectors*, using criteria that will be duly explained.

Example 10.1 Consider a two-area system, connected by a single tie-line. This system would normally have a three-dimensional state space, but if the tie-line is assumed to be perfectly reliable, the state space actually reduces to being two-dimensional. Such a system is being used for ease of graphical representation. The system and its state space are shown in Figure 10.1. This represents a hypothetical two-area power system with seven identical generators, each of capacity 100 MW, repair rate $\mu = 0.02$ per hour and failure rate $\lambda = 0.0022$ per hour (or 19.47 per year). This means that every generator has availability of 0.9. Three of the generators are in Area 1 and four are in Area 2. The two areas are connected by a tie-line of capacity 100 MW and are assumed to have constant area loads of 200 MW and 300 MW, respectively.

The tables of generation data shown in the figure describe the cumulative probability and frequency distribution functions of the generation capacities in the two areas. These distribution functions can be determined using the *unit addition algorithm* described in Chapter 8. The frequencies shown in Figure 10.1 are expressed per hour.

The complete state space can be described by the initial unclassified set U_0, which is represented as

$$U_0 = \begin{bmatrix} 300 & 400 \\ 0 & 0 \end{bmatrix}.$$

This representation consists of two boundary vectors,

(0 0) and (300 400),

which define, respectively, the bottom left corner and the top right corner of the state space shown in Figure 10.1. These boundary vectors will henceforth be referred to as the lower and upper bounds of a set, and all other members of the set, lying within the closed rectangle with the given corner points, will be referred to as states between the lower and upper bounds. In the case of a state

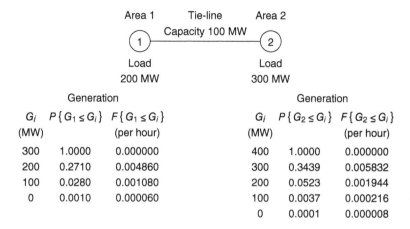

Area 1				Area 2		

Area 1   Tie-line   Area 2

Capacity 100 MW

Load   200 MW

Load   300 MW

Generation				Generation		
G_i (MW)	$P\{G_1 \le G_i\}$	$F\{G_1 \le G_i\}$ (per hour)		G_i (MW)	$P\{G_2 \le G_i\}$	$F\{G_2 \le G_i\}$ (per hour)
300	1.0000	0.000000		400	1.0000	0.000000
200	0.2710	0.004860		300	0.3439	0.005832
100	0.0280	0.001080		200	0.0523	0.001944
0	0.0010	0.000060		100	0.0037	0.000216
				0	0.0001	0.000008

(a) Two-Area System

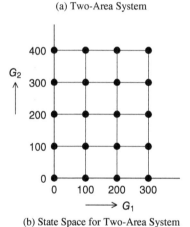

(b) State Space for Two-Area System

Figure 10.1 Two-area system and state space thereof.

space with a larger dimensionality, any set can be similarly defined as a closed hypercuboid, described by two boundary vectors.

In decomposing U_0, two partitioning vectors are defined:

1. *The lower bound of acceptability, u:*
    Suppose all the capacities are set equal to the upper boundaries of the U-set, and based on these constraints the system load curtailments are minimized; then the combination of the lowest capacity states that yield zero system curtailment constitutes the u-vector. In other words, all states between and including the u-vector and the upper boundary of the U-set constitute an A-set.

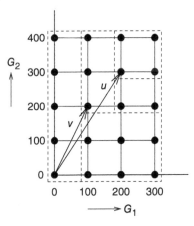

Figure 10.2 The partition vectors.

2. *The vector of minimum responsibility, v:*

Suppose all the capacity states, except the ith, are set equal to the corresponding elements of the upper boundary of the U-set; then the lowest value of the ith member that yields zero system curtailment is the ith component of the v-vector. The v-vector therefore comprises the minimum responsibilities of all the members.

For the present example, for decomposing U_0, u and v are easily determined to be:

$$u = (200\ 300)$$
$$v = (100\ 200).$$

In Figure 10.2, u and v have been plotted on the state space. The broken lines mark the boundaries of the sets resulting from the decomposition of U_0. The resulting sets are clearly shown and labeled in Figure 10.3.

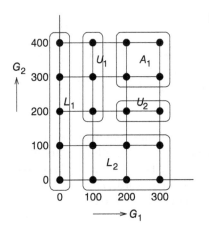

Figure 10.3 Decomposition level 1.

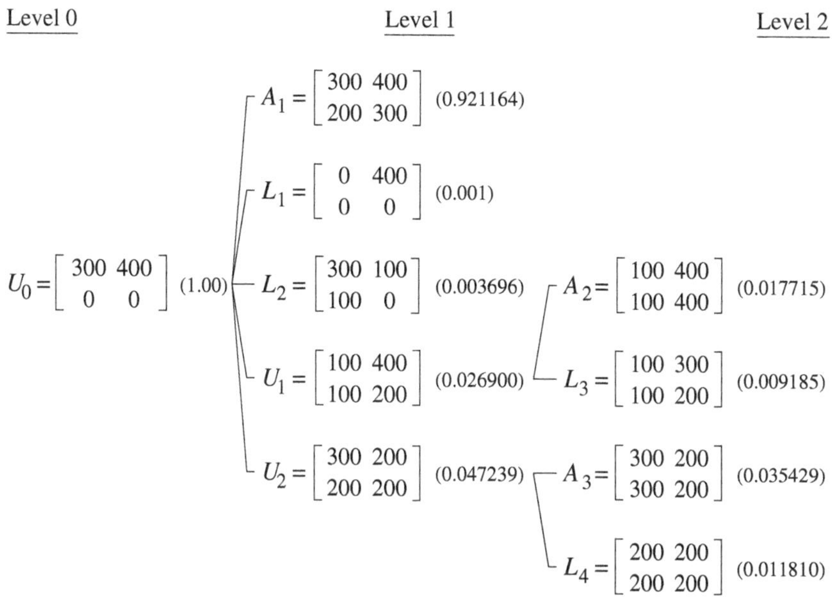

Level 0   Level 1   Level 2

$$A_1 = \begin{bmatrix} 300 & 400 \\ 200 & 300 \end{bmatrix} \ (0.921164)$$

$$L_1 = \begin{bmatrix} 0 & 400 \\ 0 & 0 \end{bmatrix} \ (0.001)$$

$$U_0 = \begin{bmatrix} 300 & 400 \\ 0 & 0 \end{bmatrix} \ (1.00) \quad L_2 = \begin{bmatrix} 300 & 100 \\ 100 & 0 \end{bmatrix} \ (0.003696) \quad A_2 = \begin{bmatrix} 100 & 400 \\ 100 & 400 \end{bmatrix} \ (0.017715)$$

$$U_1 = \begin{bmatrix} 100 & 400 \\ 100 & 200 \end{bmatrix} \ (0.026900) \quad L_3 = \begin{bmatrix} 100 & 300 \\ 100 & 200 \end{bmatrix} \ (0.009185)$$

$$U_2 = \begin{bmatrix} 300 & 200 \\ 200 & 200 \end{bmatrix} \ (0.047239) \quad A_3 = \begin{bmatrix} 300 & 200 \\ 300 & 200 \end{bmatrix} \ (0.035429)$$

$$L_4 = \begin{bmatrix} 200 & 200 \\ 200 & 200 \end{bmatrix} \ (0.011810)$$

Figure 10.4 Complete decomposition for Example 10.1.

Figure 10.3 graphically illustrates the first level of decomposition. Notice that the decomposition of U_0 resulted in the generation of one A-set, but multiple U-sets and L-sets. Even though the U-sets are connected (see Figure 10.2), they had to be defined as two disjoint sets because of the necessity to describe the sets in the form of rectangles. The same reasoning applies to the multiple L-sets.

The sets U_1 and U_2 generated from the first level of decomposition can similarly be subjected to further decomposition. For this example, the second level of decomposition turns out to be the last, since no further U-sets are produced. Figure 10.4 shows the complete decomposition for this problem. The number in parentheses next to each set gives the probability of the corresponding set.

Figure 10.4 enables the computation of the system LOLP we will calculate LOLF later):

$$\begin{aligned} \text{LOLP} &= P\{L_1\} + P\{L_2\} + P\{L_3\} + P\{L_4\} \\ &= 0.001 + 0.003696 + 0.009185 + 0.011810 \\ &= 0.025692. \end{aligned}$$

For determination of area LOLPs, the four L-sets have to be further decomposed into identical area loss of load sets (B-sets). For the decomposition of each L-set, a w-vector has to be determined. The w-vector is interpreted as follows.

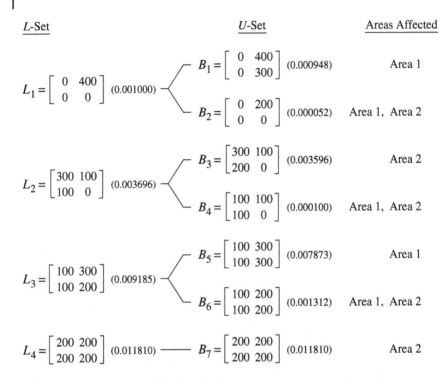

Figure 10.5 Failure set decomposition for Example 10.1.

Setting all capacity levels equal to the corresponding elements of the upper boundary of the L-set, the system curtailment is minimized. The minimum curtailment cannot be zero, but the lowest capacity levels yielding the minimum curtailment constitute the w-vector. All states between the w-vector and the upper boundary of the L-set constitute a B-set; the remaining states form W-sets, which are further decomposed in the same manner as L-sets until only B-sets remain.

Figure 10.5 shows the B-sets obtained from decomposition of the L-sets. This enables determination of the area LOLPs:

$$\text{LOLP}_1 = P\{B_1\} + P\{B_2\} + P\{B_4\} + P\{B_5\} + P\{B_6\}$$
$$= 0.000948 + 0.000052 + 0.0001 + 0.007873 + 0.001312$$
$$= 0.010285$$
$$\text{LOLP}_2 = P\{B_2\} + P\{B_3\} + P\{B_4\} + P\{B_6\} + P\{B_7\}$$
$$= 0.000052 + 0.003596 + 0.0001 + 0.001312 + 0.011810$$
$$= 0.016870.$$

It is appropriate to remark that a realistic system would have a state space of much larger size and dimensionality, and the number of decompositions required would be substantially large. This is exacerbated when multiple load states are used, as we shall see in the next section. Section 10.4 discusses this problem further, and proposes a practical solution.

10.3.2   Simultaneous Decomposition

The method of decomposition described above applies to one given load scenario. To perform this for every hour of forecasted load, or even for daily peaks, would be computationally intractable. This problem is circumvented by using the cluster load model described in Section 9.3.3.

The obvious application of the cluster model is in the form of the *extended decomposition* method [39], where decomposition is performed for every load cluster, and the computed indices are weighted by the corresponding cluster probabilities and aggregated. This can be made considerably more efficient, with no loss of accuracy, by using the *simultaneous decomposition* method [56], described here.

The generation and load models are so constructed that the states are integral multiples of the same increment. Then a reference load state is defined, using which the generation model is modified for every load cluster, and these modified models are interleaved to construct what will be referred to as the *integrated generation model*. When the integrated generation model is used to represent the area generation in the $(N_n + N_a)$-dimensional state space described in Section 9.3.8, the resulting state space is referred to as the *integrated state space.* Subjecting the integrated state space to decomposition with the reference load state as the load constitutes the method of Simultaneous Decomposition.

Using an identical concept, planned outages of generators can also be taken into consideration [56]. An integrated generation model can be constructed for every maintenance interval, and these can be interleaved together to form the integrated generation model over the entire study period.

The Integrated State Space
From the cluster load models over all the maintenance intervals, an appropriate load model is chosen for every area to constitute the reference load level vector. The same reference load vector must serve as reference over the entire period under consideration, so that all the models can be interleaved into a single integrated state space [56].

For a given maintenance interval, the corresponding generation model in every area is modified for every cluster load level in that interval. This modification consists of "shifting" a generation model by a capacity level equal to the difference between the cluster load and the reference load for that area, so that the margin between any generation level in the modified model and the reference load remains the same as that between the corresponding capacity level in the original model and the cluster load. These modified models are interleaved together to form an integrated generation model for the given maintenance interval. This is repeated for every maintenance interval, and the resulting models are further interleaved and combined with the transmission model to produce the integrated state space.

The above operations result in a many-to-one mapping of the states in the original state space onto the integrated state space. If there are N_i maintenance intervals and N_c cluster load levels, then, in the limiting case, as many as $N_i \times N_c$ states in the original state space can map onto a single state in the integrated state space. This "condensing" of the state space is what makes the simultaneous decomposition algorithm so efficient.

It is necessary to preserve the original generation models because of the necessity to refer back to these models when computing the indices. Before proceeding to the methodology for computing reliability indices from simultaneous decomposition, it is appropriate to illustrate, by means of an example, the construction of an integrated state space.

Example 10.2   Consider the two-area system of Example 10.1, with two load states:

$$\mathbf{d}^1 = (200\ 300), \quad P\{\mathbf{D} = \mathbf{d}^1\} = 0.5$$
$$\mathbf{d}^2 = (100\ 200), \quad P\{\mathbf{D} = \mathbf{d}^2\} = 0.5.$$

This system may be represented as shown in Figure 10.6(a). The state space for this system is in fact the same as that shown in Figure 10.1. However, in applying the method of simultaneous decomposition, a reference load state must first be selected. It is convenient to select the highest load state as reference. The reason for this will be explained later. So state 1 serves as the reference. For this (reference) state, the generation model and the state space are left unaltered. For state 2, however, the state space requires modification. The system representation corresponding to state 2 is as shown in Figure 10.6(b). The modification of this model, as stated earlier, consists of "shifting" it by a capacity level equal to the difference between the cluster load for state 2 and the reference load, so that the margin between any generation level in the modified model and the reference load remains the same as that between the corresponding capacity level in the original model and the cluster load for state 2. This results in the modified model shown in Figure 10.6(c).

Observe that the modified model of Figure 10.6(c) can be reasonably expected to behave in the same manner as the original model of Figure 10.6(b), yet the load level is the same as the reference load. The modified state space for this model is shown in Figure 10.7(b); Figure 10.7(a) shows the original state

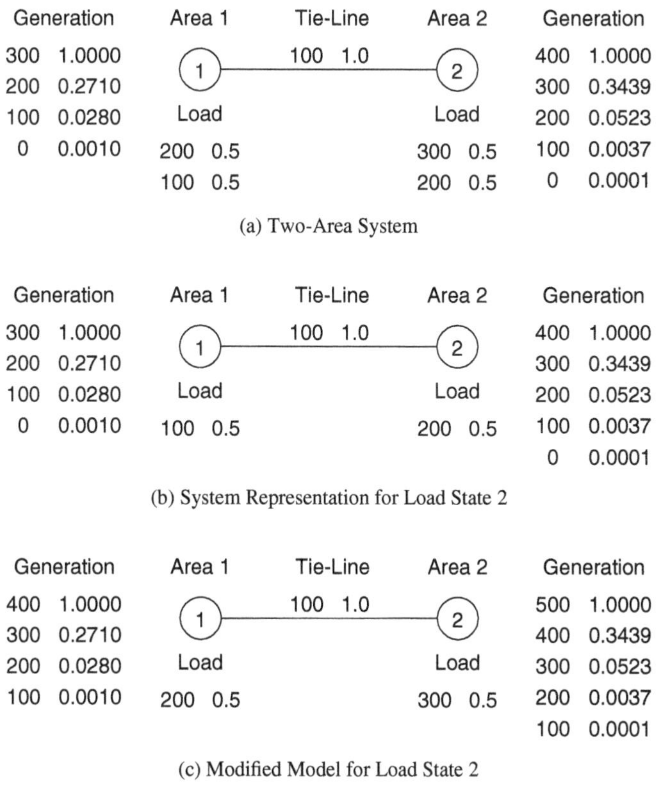

Generation	Area 1	Tie-Line	Area 2	Generation
300 1.0000		100 1.0		400 1.0000
200 0.2710	①————————②			300 0.3439
100 0.0280	Load		Load	200 0.0523
0 0.0010	200 0.5		300 0.5	100 0.0037
	100 0.5		200 0.5	0 0.0001

(a) Two-Area System

Generation	Area 1	Tie-Line	Area 2	Generation
300 1.0000		100 1.0		400 1.0000
200 0.2710	①————————②			300 0.3439
100 0.0280	Load		Load	200 0.0523
0 0.0010	100 0.5		200 0.5	100 0.0037
				0 0.0001

(b) System Representation for Load State 2

Generation	Area 1	Tie-Line	Area 2	Generation
400 1.0000		100 1.0		500 1.0000
300 0.2710	①————————②			400 0.3439
200 0.0280	Load		Load	300 0.0523
100 0.0010	200 0.5		300 0.5	200 0.0037
				100 0.0001

(c) Modified Model for Load State 2

Figure 10.6 Modification of generation models.

space, corresponding to load state 1. Since both these spaces pertain to the same (reference) load state, they may be superimposed into a single integrated state space, as shown in Figure 10.7(c), and the integrated state space may be subjected to decomposition.

Figure 10.7 also makes it evident why it is convenient to use the highest load level as reference: all points in the integrated state space remain nonnegative.

10.3.3   Computation of Indices

It has been stated above that each state in the integrated state space can be mapped back to as many as $N_i \times N_c$ pre-images in the original state space. The same argument applies to sets of states. So if the probability of a set in the integrated state space is to be computed, then the set must first be mapped back to all its pre-images in the original state space, and the sum of the probabilities of all these pre-image sets must be computed to yield the probability of the integrated set.

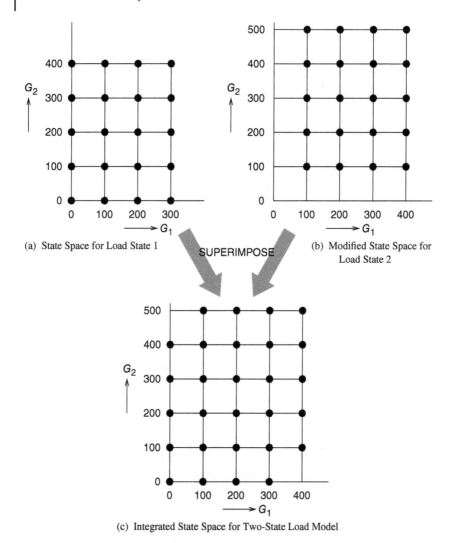

(a) State Space for Load State 1

SUPERIMPOSE

(b) Modified State Space for Load State 2

(c) Integrated State Space for Two-State Load Model

Figure 10.7 Integrated state space for Example 10.2.

The sum of the probabilities of the L-sets yields the system LOLP, and the sum of the probabilities of the B-sets corresponding to each area loss of load gives us the area for that area. Determination of the area EUEs proceeds as follows. For every integrated B-set, all the pre-image sets are determined. For each pre-image set, the mean generation vector is determined. From the mean generation vector and the actual load vector for the pre-image set, the mean curtailment vector is determined. The product of this curtailment vector and the probability of the pre-image set yields the contribution of this pre-image set towards the area EUEs. The sum of the contributions of all the pre-image sets is the total contribution of the integrated B-set. The total locational (area or bus) EUEs are computed from all the B-sets. The system EUE is the sum of the locational EUEs.

So far we have presented the calculation of LOLP in an intuitive manner. We shall now formalize the method for calculation of LOLP and LOLF. First we shall develop the theory. Then we shall show how the theoretical approach is implemented. Finally, we shall illustrate it with an example.

Theoretical Approach

For the purpose of the forthcoming analysis, we shall use the following assumptions:

(a) The system is frequency balanced [7], i.e., the frequency between any pair of states is equal in both directions. This automatically holds true if every component can be represented by a two-state Markov model. In cases where one or more components assume multiple states, frequency balance does not always apply. However, in such cases, forced frequency balance can be imposed. This is achieved by the following device. For every downward transition, say from system state k to state m, by degradation of the rth component from component state s to state t, the actual downward transition rate λ_{st} is replaced by the fictitious transition rate λ'_{st}, given by [7]:

$$\lambda'_{st} = \frac{p_t}{p_s} \lambda_{ts}, \tag{10.1}$$

where p_t = probability of component r assuming state t, and λ_{ts} = actual upward transition rate of component r from state t to state s. In the next section we will describe a method that automatically ensures frequency balance.

(b) The system can transit in one step from one state to another only by failure or restoration of a single component.

A transition to a higher capacity state is said to be an upward transition; similarly, a transition to a lower capacity state is said to be a downward transition.

Assume that decomposition has resulted in N_L failure sets; then the aggregate failure set is given by:

$$L = \bigcup_{i=1}^{N_L} L_i. \tag{10.2}$$

Clearly, since the sets are all disjoint, the failure probability is

$$P\{L\} = \sum_{i=1}^{N_L} P\{L_i\}. \tag{10.3}$$

The following analysis is a generalization of the results developed in [7]. Consider the sums described by (10.4) and (10.5), given below:

$$F^+\{L_i\} = \sum_{x \in L_i} \left[P\{x\} \sum_{j=1}^{N_C} \lambda_{xj}^+ \right] \tag{10.4}$$

$$F^-\{L_i\} = \sum_{x \in L_i} \left[P\{x\} \sum_{j=1}^{N_C} \lambda_{xj}^- \right], \tag{10.5}$$

where

λ_{xj}^+: transition rate of component j from its state in system state x to higher-capacity states;

λ_{xj}^-: transition rate of component j from its state in system state x to lower-capacity states.

$F^+\{L_i\}$ is the sum of frequencies of all upward transitions from all the states in the subset L_i. Similarly $F^-\{L_i\}$ is the sum of frequencies of all downward transitions from all the states in the subset L_i. We will use this notation for the remainder of the analysis.

Now consider a state x in set L_i. The sum of frequencies of upward transitions from this state is

$$F^+\{L_i(x)\} = P\{x\} \sum_{j=1}^{N_C} \lambda_{xj}^+, \tag{10.6}$$

and similarly, the downward frequency is

$$F^-\{L_i(x)\} = P\{x\} \sum_{j=1}^{N_C} \lambda_{xj}^-. \tag{10.7}$$

Since the system is considered coherent, the $F^-\{L_i(x)\}$ transitions cannot cross the boundary between L-sets and A-sets. The upward transitions can be split into two components,

$$F^+\{L_i(x)\} = F_0^+\{L_i(x)\} + F_1^+\{L_i(x)\}, \tag{10.8}$$

such that the $F_0^+\{L_i(x)\}$ component is the one that crosses the boundary between the L-sets and A-sets, and $F_1^+\{L_i(x)\}$ remains within the L-set. Since the components are assumed independent and are frequency balanced for any two states x_i and x_j,

$$F\{x_i \rightarrow x_j\} = F\{x_j \rightarrow x_i\}. \tag{10.9}$$

Now if we consider

$$F\{L\} = \sum_{i=1}^{N_L} \left[F^+\{L_i\} - F^-\{L_i\} \right], \tag{10.10}$$

$F\{L\}$ can be expressed as follows:

$$F\{L\} = \sum_{i=1}^{N_L} \sum_{x \in L_i} \left[F_0^+\{L_i(x)\} + F_1^+\{L_i(x)\} - F^-\{L_i(x)\} \right]. \tag{10.11}$$

Because of (10.9), we have

$$\sum_{i=1}^{N_L} \sum_{x \in L_i} F_1^+\{L_i(x)\} = \sum_{i=1}^{N_L} \sum_{x \in L_i} F^-\{L_i(x)\}. \tag{10.12}$$

This can be understood as follows. Starting from a failed state, an upward transition that does not cross the boundary is actually an upward transition from a failed state to another failed state. For every upward transition from a failed state x_i to another failed state x_j, there is a downward transition from x_j to x_i of equal frequency. Hence the two sums in (10.12) are equal. From (10.11) and (10.12), we have

$$F\{L\} = \sum_{i=1}^{N_L} \sum_{x \in L_i} \left[F_0^+\{L_i(x)\} \right]. \tag{10.13}$$

Equation (10.13) represents the expected transitions across the boundary of L-sets and A-sets. We have seen before that when the state space is divided into two disjoint sets, the steady state frequency is equal in both directions. Therefore the frequency from L-sets to A-sets is the same as from A-sets to L-sets. Thus (10.13) represents the failure frequency, i.e., the LOLF. Equation (10.10) is equivalent to (10.13) and is suitable for calculating the frequency of failure of the system, as shown in the next section.

Practical Implementation

Equations (11.6) and (10.10) are used to determine the probability and frequency of system failure. We will now show how the data can be constructed so that calculation of $P\{L_i\}$ and $F\{L_i\}$ can be achieved with very little effort. This can be done as follows.

Consider a system with N_C components where each component has several capacity states, of which every state has a known probability and frequency of occurrence. Construct cumulative probability and frequency distributions for each component, so that P_{kj} and F_{kj} denote, respectively, the cumulative probability and frequency of state k of component j. The states are so indexed that a higher index implies a higher capacity. For this system, a state space is constructed, and decomposed into disjoint A-sets and L-sets, using the method described in Section 10.3.

Now consider a failure set L_i. Recall that L_i is described by a maximum state and a minimum state. Assume that states n and m of component j define the maximum and minimum states of set L_i. For state k of component j, where $m \leq k \leq n$, the cumulative probability and frequency can be expressed as follows:

$$P_{kj} = P_{(k-1)j} + p_{kj}, \tag{10.14}$$

and

$$F_{kj} = F_{(k-1)j} + p_{kj}\left(\lambda_{kj}^+ - \lambda_{kj}^- \right), \tag{10.15}$$

where

p_{kj}:   probability of state k of component j;
$\lambda_{kj}^+, \lambda_{kj}^-$: transition rates from state k of component j to higher and lower capacity states, respectively.

From (10.14) and (10.15), we get

$$\lambda_{kj}^+ - \lambda_{kj}^- = \frac{\left(F_{kj} - F_{(k-1)j} \right)}{\left(P_{kj} - P_{(k-1)j} \right)}. \tag{10.16}$$

If we now represent the states from m to n of component j by an equivalent state s, then the equivalent values for this can be expressed in the same manner as (10.16):

$$\lambda_{sj}^+ - \lambda_{sj}^- = \frac{\left(F_{nj} - F_{(m-1)j} \right)}{\left(P_{nj} - P_{(m-1)j} \right)}. \tag{10.17}$$

Similarly, the probability of this equivalent state can be written as

$$p_{sj} = P_{nj} - P_{(m-1)j}. \tag{10.18}$$

In a similar fashion, for every component, the states between the maximum and minimum of set L_i can be represented by an equivalent state.

Then the probability of L_i is:

$$P\{L_i\} = \prod_{j=1}^{N_C} p_{sj}, \tag{10.19}$$

and the frequency is:

$$F^+\{L_i\} - F^-\{L_i\} = P\{L_i\} \sum_{j=1}^{N_C} \left(\lambda_{sj}^+ - \lambda_{sj}^- \right). \tag{10.20}$$

Hence the probability of system failure is determined by substituting $P\{L_i\}$ from (10.19) into (11.6), and the frequency is determined by substituting $F^+\{L_i\} - F^-\{L_i\}$ from (10.20) into (10.10), as shown below:

$$F\{L\} = \sum_{i=1}^{N_L} P\{L_i\} \sum_{j=1}^{N_C} \left(\lambda_{sj}^+ - \lambda_{sj}^- \right). \tag{10.21}$$

Notice that the determination of the failure frequency using the expression given in (10.20) adds little to the computational effort expended in decomposing the state space and calculating the failure probability using (10.19).

This approach will now be illustrated by means of an example.

Example 10.3   Consider the same two-area system that we analyzed in Example 10.1. The probability and frequency distributions and the state space are shown in Figure 10.1. Figure 10.4 shows the decomposition stages, and the decomposed state space is shown in Figure 10.8.

Table 10.1 shows the probabilities and frequencies of the four L-sets and the total probability and frequency of system failure. Transition rates and frequencies are expressed per hour. In this table, $P\{L_i\}$ is obtained using (10.19). $\Delta\lambda\{L_i\}$ is determined by

$$\Delta\lambda\{L_i\} = \sum_{j=1}^{N_C} \left(\lambda_{sj}^+ - \lambda_{sj}^- \right),$$

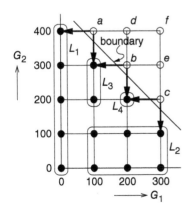

Figure 10.8 State space showing failure states and boundary.

Table 10.1 Probabilities and Frequencies of L-Sets

L_i	$P\{L_i\}$	$\Delta\lambda\{L_i\}$	$F\{L_i\}$
L_1	0.001000	0.060000	6.0000×10^{-5}
L_2	0.003696	0.058318	2.1556×10^{-4}
L_3	0.009185	0.054286	4.9864×10^{-4}
L_4	0.011810	0.051111	6.0361×10^{-4}
$\sum_i L_i$	0.025692		1.3778×10^{-3}

where the term in the parentheses is obtained using (10.17). $F\{L_i\}$ corresponds to the left hand side of (10.20). The total probability and frequency of system failure appear on the last row of Table 10.1.

Notice that for this small system, these results can easily be validated by enumeration. Since the number of A-states is less than that of L-states, it is easier to calculate the indices using A-states. The six A-states are marked a through f in Figure 10.8:

$$P\{L\} = 1 - (p_a + p_b + p_c + p_d + p_e + p_f)$$
$$= 1 - (0.017715 + 0.070859 + 0.035429$$
$$+ 0.159432 + 0.212576 + 0.478297)$$
$$= 0.025692$$
$$F\{L\} = (p_a + p_b + p_c)5\lambda$$
$$= 0.124003 \times 0.011111\dot{1}$$
$$= 1.3778 \times 10^{-3} \text{ per hour.}$$

The terms $p_a, p_b, \ldots$ denote exact state probabilities. The 5λ term is because each of states a, b and c corresponds to a total of five generators available, so the total downward transition rate from each of these states is 5λ.

It is important to note that in most practical systems the state space is much larger, and the decomposition needs to be performed using a computer program.

10.3.4 Concluding Remarks concerning State Space Decomposition

As mentioned earlier, decomposition can prove to be a very effective method because it deals with sets of states rather than individual states. However, in practice it has been found that as the process proceeds through increasingly higher levels (recursions), its yield diminishes, i.e., the number of sets produced at each level increases, but the total probability of the states classified at each level decreases. Consequently, beyond a certain point, the effort per recursion increases but the contribution of each recursion diminishes. The increasing effort per recursion results from (a) increasing number of sets; (b) increasing effort in identifying partition vectors for each unclassified set; and (c) increasing book keeping effort, i.e., effort required to keep track of all the sets generated and their probability and frequency contributions to the reliability indices.

One practical solution to this is to discard unclassified sets whose probabilities fall below a certain (suitably small) threshold. This results in early termination of the process, without achieving complete classification, without significant error. Another approach is to combine decomposition with Monte Carlo simulation, as described earlier in Section 10.2.6.

10.4 Conclusion

The description of the method of state space decomposition illustrated some of the considerations that go into the reliability analysis of interconnected systems. As stated early in the chapter, there are several methods in existence, and the choice of method is determined by such factors as the system topology, the depth (or order of contingencies) that must be considered, complexity of the system and the operating policies that must be modeled and the amount of accuracy that is desired. This chapter has provided ample references to enable to reader to explore any of the listed methods in greater detail.

In concluding this chapter, it is pointed out again, as done earlier in the text, that considerable engineering judgment goes into the modeling and analysis of the reliability of large power systems.

Exercises

10.1   Consider a capacity-constrained network, as shown in Figure 10.9. The capacity of each sending arc is represented by the state-transition diagram shown in Figure 10.10. Each transmission arc (shown by a broken line) has a capacity of 100 MW in each direction and each receiving arc has a capacity of 100 MW. Assume that $\lambda = 0.2$ failures/h, $\mu = 2$ repairs/h.

Manually perform a state space decomposition of this system and determine the probability and frequency of system failure.

Figure 10.9 Three-node network.

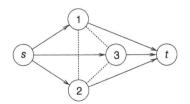

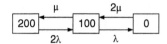

Figure 10.10 State-transition diagram for each sending arc.

10.2 Consider a three-area system where the area generations have capacities and probabilities as given below.

Area 1		Area 2		Area 3	
capacity (MW)	cumulative probability	capacity (MW)	cumulative probability	capacity (MW)	cumulative probability
		600	1.0		
500	1.0	500	0.737856	500	1.0
400	0.67232	400	0.344640	400	0.67232
300	0.26272	300	0.098880	300	0.26272
200	0.05792	200	0.016960	200	0.05792
100	0.00672	100	0.001600	100	0.00672
0	0.00032	0	0.000064	0	0.00032

Each transmission line has two states, 100 MW and 0 MW, with probabilities 0.99 and 0.01, respectively. The area loads are 400 MW, 500 MW and 400 MW in Areas 1, 2 and 3, respectively. Find the probability of loss of load in the system.

11

Reliability Evaluation of Composite Power Systems

11.1 Introduction

Composite power system reliability evaluation is concerned with the ability of the combined generation and transmission system to supply adequate and suitable electrical energy to major system load centers. Whereas in multi-area reliability evaluation the transmission constraints are only indirectly considered, in composite power system reliability evaluation the internal transmission limitations are directly modeled. These reliability studies can help better represent generation effects in transmission system reliability analysis, optimize relative investments in generation and transmission systems, and include dispersed generation. As stated in Chapter 9, composite system reliability evaluation is performed using analytical, Monte Carlo simulation or hybrid methods.

11.2 Analytical Methods

Most analytical methods are based on state space enumeration. However, due to the size and complexity of composite power systems, exhaustive enumeration is often impractical. For this reason, several methods have been developed to identify as large a part of the set of failure states as possible. *Contingency ranking and selection* is one of the techniques that has been used in a program developed by the Electric Power Research Institute. The goal of contingency selection techniques is to determine from the set of all possible contingencies the subset that will cause system failure. No contingency selection method can attain this goal perfectly; they can at best provide a subset that contains most contingencies causing system failure.

One approach is to rank contingencies by first solving each contingency using DC load flow, but it would be very time consuming. In a faster but less accurate method contingencies are ranked approximately by severity based on a performance index. The scalar function, called performance index (PI) is first

Electric Power Grid Reliability Evaluation: Models and Methods, First Edition. Chanan Singh,
Panida Jirutitijaroen, and Joydeep Mitra.
© 2019 by The Institute of Electrical and Electronic Engineers, Inc. Published 2019 by John Wiley & Sons, Inc.

defined to provide a measure of system stress. Then a suitable technique is used for predicting ΔPI, i.e., the change in PI when a component is outaged. The ΔPI values for contingencies are then used to rank them in the order of severity. Then AC or DC load flows are used to determine which of these ranked contingencies actually do cause problems. When a certain specified number of consecutive contingencies do not lead to system failure, the process is stopped. The assumption here is that remaining lower-ranked contingencies will also not cause system failure. This is not a foolproof method of ranking contingencies. It is possible that some severe contingencies may be left out and, also some not so severe contingencies may be ranked. Contingency ranking may be done either based on overload or voltage problems. A brief discussion is provided here for contingency ranking based on overloads. A performance index used for this purpose is [58]:

$$\text{PI} = \sum_{\ell} W_{\ell} \left(\frac{P_{\ell}}{P_{\ell}^{\max}} \right)^n, \tag{11.1}$$

where

W_{ℓ}:   weighting factor for circuit ℓ;
P_{ℓ}:   real power flow on circuit ℓ;
$P_{\ell}^{\max}$:   power rating of circuit ℓ;
n: an even integer, generally 2.

Several approaches for finding ΔPI have been proposed. PI can generally provide a good measure of system stress. In cases where load in one branch increases but in others decreases, the PI may fail to recognize overloads, resulting in a masking phenomenon. Masking can be reduced by increasing the exponent n, but it becomes difficult to solve ΔPI for $n > 2$.

11.2.1   Performance Index for Generator Outages

Generator outages are ranked based on the prediction of overloads resulting from outages. The PI used is the same as that for lines. The change in PI resulting from a change in P_i, the power injection at the ith bus, is estimated by:

$$\Delta \text{PI} = \frac{\partial \text{PI}}{\partial P_i} \Delta P_i. \tag{11.2}$$

This linear predictor has been found to produce reasonably good rankings for generating unit outages. An implementation using DC power flow is shown below:

$$\frac{\partial \text{PI}}{\partial P_i} = n\hat{\theta}_i,$$

where

$\hat{\theta}_i = i$th component of vector $\hat{\theta}$;

$N = $ number of buses;

$\hat{\theta} = B^{-1}\hat{P}$;

$B = N \times N$ system susceptance matrix;

$\hat{P} = $ vector of length N built by adding for every circuit ℓ an injection of $\hat{P}_\ell$ on the sending end bus and $-\hat{P}_\ell$ on the receiving end bus of circuit ℓ, where

$$\hat{P}_\ell = \frac{W_\ell B_\ell^n \theta_\ell^{n-1}}{\bar{P}_\ell^n}. \tag{11.3}$$

In other words, $\hat{P}$ is the sum of N_ℓ vectors (where N_ℓ is the number of circuits), each of which contains $\hat{P}_\ell$ on the sending end bus and $-\hat{P}_\ell$ on the receiving end bus of circuit ℓ, and 0 at all other buses. In (11.3), θ_ℓ denotes the angle difference between the sending end bus bus and receiving end bus of circuit ℓ. After the base case DC flow calculation, elements of B and the values of θ_ℓ are available. One additional calculation is required to find $\hat{\theta}$.

11.2.2 Performance Index for Voltage Deviations

When a circuit is outaged, two factors contribute to voltage drop: loss of charging of the outaged circuit and increased VAr consumption on circuits that have increased loading as a result of this outage. One of the simplest PI is:

$$PI = \sum_\ell X_\ell P_\ell^2,$$

where X_ℓ is the reactance of circuit ℓ. A performance index that has shown more success is:

$$PI = \sum_\ell X_\ell \left(\frac{1}{P_{oi}^2} + \frac{1}{P_{oj}^2} \right)^{0.25} P_\ell^2,$$

where

$X_\ell = $ reactance of circuit ℓ;

$P_\ell = $ power flow on circuit ℓ;

$i, j = $ sending and receiving end buses of circuit ℓ;

$P_{oi}, P_{oj} = $ terms recognizing line charging and/or reactive sources and loads on buses i and j.

11.2.3 Selection of Contingencies

The contingencies are evaluated in decreasing order of severity. For each single-order contingency, secondary contingencies can also be ranked and evaluated. Evaluation is stopped either if a prespecified number of successes are encountered or if the contingency probability is lower than a threshold.

11.3 Monte Carlo Simulation

In Chapter 6 we described general approaches that can be used to estimate the reliability of most systems. As mentioned in Chapter 6, it is appropriate to apply Monte Carlo methods in situations where analytical methods are difficult or impossible to apply, for reasons such as system or model complexity, or the necessity to capture probabilistic or temporal dependencies in system transitions. This is true of power systems, and Monte Carlo simulation has been applied in several forms to assess the reliability of power systems. Several variants can be found in the literature. Some of the models and methods are described in this chapter.

Both sequential and nonsequential Monte Carlo methods have been applied to composite system reliability assessment, and Monte Carlo methods probably constitute the most widely used methods for composite system studies. These methods, as applied to composite systems, are described here in detail. The choice of whether to use sequential or nonsequential simulation is determined by the specific needs of the study. If it is necessary to consider dependent events and temporal relationships, then sequential simulation is necessary; if these dependencies do not exist or can be considered negligible, then nonsequential simulation is often preferred because it generally converges faster than sequential methods.

11.4 Sequential Simulation

This is appropriate in systems where it is necessary to model dependent events and temporal relationships, such as the effects of start-up, shut-down and ramping times, costs, etc. A model is built that incorporates the necessary amount of detail in system operation and dependencies, and the simulation is executed for the period of interest. Each such "run" constitutes a *sample path* (also known as *realization* or *replication*). Over each sample path, data is collected for the estimation of the desired indices. This data could consist of the frequency of occurrence of system failure events, total and fractional time spent in these states and the amount of power and energy not served as a result of system failures.

The data thus collected may be used to estimate system reliability indices such as *Loss of Load Probability* (LOLP), *Loss of Load Frequency* (LOLF), *Mean Down Time* (MDT), *Mean Time Between Failures* (MTBF), *Expected Power Not Served* (EPNS) and *Expected Unserved Energy* (EUE, also known as EENS). In addition to system indices, one may be interested in locational indices such as *Bus* LOLP, LOLF, EUE and EPNS at one or more buses. It is necessary to execute simulations over several (sometimes hundreds or even thousands) sample paths before the estimates converge to acceptable tolerances.

As described in Chapter 6, there are three predominant variants of sequential simulation that are used in power system reliability assessment—synchronous timing, asynchronous timing and mixed timing. The synchronous timing method is appropriate in systems where changes occur in the system at regular intervals. An example is a system where the load changes hourly. The asynchronous timing method is appropriate in systems where changes do not occur at regular time intervals. An example is an emergency or standby system. Reference [59] describes an application of asynchronous sequential simulation to a standby system.

Mixed timing incorporates a combination of both synchronous and asynchronous timing controls. This is very commonly used in the industry for purposes of both reliability evaluation and production simulation. In its most common form, it consists of traversing an hourly load curve (historical or forecast) over the period of interest, and advancing the states of system components asynchronously. The following algorithm describes a typical embodiment of this method:

1. Input system data (component state capacities and transition rates, dependencies, interconnection information, maintenance schedules and hourly load curve). Define convergence criteria for indices.
2. Initialize sample path index at $i = 1$.
3. Initialize component states at hour $j = 1$.
4. For prevailing component states at hour j, perform the following: For every component, draw a random number, and determine the time to the next transition using one of the methods outlined in Section 6.3.1. Of these times, select the smallest, $t_{min}(k_m)$; the corresponding time and transition indicate the most imminent event, i.e., after t_{min}, the component k will transit to state m.
5. At hour j, perform the following:
   (a) Decrease by one hour the time t_{min} to the next event as determined in step 4.
      (i) If $t_{min} > 0$, no change has occurred in system states; proceed to step 5(b).
      (ii) If $t_{min} = 0$, update state of component k to m for hour j. If there are dependencies triggered by this transition, update states accordingly.

For prevailing component states at hour j, perform the following: For every component, draw a random number and determine the time to the next transition. Of these times, select the smallest, $t_{min}(k_m)$. In other words, determine a new $t_{min}(k_m)$ as soon as a transition occurs.

(iii) At hour j, is any component taken out for maintenance or brought back into service after maintenance? If so, alter its state accordingly.

(b) For the prevailing component states as determined in step 5(a), and the load at hour j, determine if the system experiences loss of load. If so, update estimates of indices as described in Section 6.4.2; otherwise proceed to step 6.

6. Is j equal to the last hour in study period? If so, proceed to step 7; otherwise increase j by one hour and go to step 5.

7. Have the estimates of the indices converged to criteria specified in step 1? If so, end simulation and report results (index estimates, and, optionally, convergence related information, computation time, etc.); otherwise increase sample path index i by 1 and go to step 3.

This algorithm serves to convey an idea of the mechanics of the mixed timing sequential simulation method. In general, for an interconnected power system, there may be several hourly load curves for different buses in the system, and these will have to be suitably accommodated in step 5(b). Moreover, there may be variations of this method. For instance, instead of considering only one imminent event, it is possible to maintain, and initialize at step 4, a different "clock" for each component, which shows the time to the next transition for that component. Every component "clock" would then be updated every hour and reset when the component changes states. Other variations are also possible.

As stated before, this method is very commonly used in the industry. It is important to be aware that in several commercial embodiments of this method, the test for convergence is not effectively implemented and provides the user with the option of executing the simulation over only one sample path, without warning the user that any statistical inference drawn from one sample path is almost guaranteed to be grossly inaccurate. It is a good idea for the educated user to bear this in mind and avoid the temptation of executing only one replication of what is often a very time-consuming simulation.

In step 5(b), a testing of system performance is involved. Different methods for testing of system states are discussed in Section 11.6.

11.4.1 Estimation of Indices

The estimation of reliability indices from sequential simulation is based on the concepts described in Section 6.4.2. In this section we show how indices specific to bulk power reliability are estimated.

Mean down time (MDT)

The MDT is also known as the average interruption duration, and is often denoted by $\bar{r}$. By definition,

$$\text{MDT} = \bar{r} = \int_0^\infty r f_R(r) dr, \tag{11.4}$$

where R is the random variable denoting the system down time, and $f_R(r)$ is the probability density function of R. From sequential simulation, MDT is estimated using:

$$\text{MDT} = E[\hat{r}]; \quad \hat{r} = \frac{1}{N^{\text{cy}}} \sum_{i=1}^{N^{\text{cy}}} T_i^{\text{dn}}, \tag{11.5}$$

where $E[\cdot]$ is the expectation operator, $\hat{r}$ is the estimator of MDT, T_i^{dn} is the duration of ith interruption encountered during the sequential simulation and N^{cy} is the number of *cycles* simulated. A cycle consists of a service period T_i^{up} and an interruption period T_i^{dn}; the ith *cycle time* T_i^{cy} equals $T_i^{\text{up}} + T_i^{\text{dn}}$. The total period of simulation T is given by $T = N^{\text{cy}} T^{\text{cy}} = N^{\text{cy}} \left(T_i^{\text{up}} + T_i^{\text{dn}} \right)$. The total time or simulation, or the number of cycles that are simulated, is dependent on the desired convergence criterion, as described in Section 11.4.2.

Loss of load probability (LOLP)

The LOLP index can be estimated as follows:

$$\text{LOLP} = E[\Pi]; \quad \Pi = \frac{1}{T} \sum_{i=1}^{N^{\text{cy}}} T_i^{\text{dn}}, \tag{11.6}$$

where Π is the estimator of LOLP, and the other variables are as defined above.

Loss of load frequency (LOLF)

The LOLF index gives the frequency of service outage and can be estimated as follows:

$$\text{LOLF} = E[\Phi]; \quad \Phi = \frac{N^{\text{cy}}}{T}, \tag{11.7}$$

where Φ is the estimator of LOLF and the other variables are as defined above. It should be noted that the unit of LOLF is failures per unit time; hence, if time is tracked in hours, (11.7) will yield LOLF in f/h, and may need to be converted into f/y, which is the customary unit for expressing LOLF. Also, LOLF is related to LOLP and MDT as follows:

$$\text{LOLF} = \frac{\text{LOLP}}{\text{MDT}}. \tag{11.8}$$

Expected power not served (EPNS)

The EPNS index, also known as the expected demand not supplied (EDNS), is the sum of the products of probabilities of failure states and the corresponding load curtailments, which can be estimated as follows:

$$\text{EPNS} = \sum_{x_i \in X_f} P\{x_i\} \times C\{x_i\}, \tag{11.9}$$

where $P\{x_i\}$ and $C\{x_i\}$ are the probability of occurrence of state x_i, and the system load curtailment in state x_i, and X_f is the set of failure states. Using sequential simulation, EPNS is estimated from:

$$\text{EPNS} = E[\Psi]; \quad \Psi = \frac{1}{T} \sum_{i=1}^{N_C} T_i^{\text{dn}} C\{x_i\}, \tag{11.10}$$

where Ψ is the estimator of EPNS, and $C\{x_i\}$ is the system load curtailment in the prevailing state.

11.4.2 Stopping Criterion

As mentioned in Section 6.4.2, a stopping criterion is applied to stop the algorithm when the reliability indices converge. This determines the number of cycles that are simulated. The stopping criterion is applied as follows:

$$\eta = \frac{\sqrt{Var(\rho_{N^{\text{cy}}})}}{E\left[\rho_{N^{\text{cy}}}\right]} \leq \varepsilon, \tag{11.11}$$

where η is the coefficient of variation, $Var(\cdot)$ is the variance function, $\rho_{N^{\text{cy}}}$ is the value of the estimate of the reliability index of interest (such as LOLP or EPNS) at the end of N^{cy} cycles and ε is a prespecified tolerance.

At intervals of several cycles, η is calculated. If this value is less than or equal to the specified tolerance ε, the algorithm is terminated; otherwise, the simulation continues.

11.5 Nonsequential Simulation

This is appropriate in systems where component and temporal dependencies either do not exist or can be neglected with acceptable error. As stated before, the aggregate of states that the system can assume during the period of interest is treated as the population out of which a sample of appropriate size is drawn, and population indices are estimated from sample statistics.

11.5.1 Algorithm

A typical implementation of nonsequential Monte Carlo simulation is described in the following algorithm. This implementation uses discrete probability distribution functions as described in Section 6.3.1. Random sampling is performed on these distributions in the manner described in Section 6.3.1; the algorithm below describes how it is applied specifically for composite system evaluation.

1. Input system data (component state capacities and probabilities, interconnection information, maintenance schedules and load states). Define convergence criteria for indices.
2. Based on the maintenance schedules, divide the study period into intervals such that one or more components are taken out for maintenance or brought back from maintenance only at the beginning or end of an interval. In other words, the only outages that occur *within* an interval are forced outages, *not* planned outages. Based on the relative durations of the maintenance intervals, construct a discrete probability distribution function over all the maintenance intervals.
3. For every maintenance interval, arrange the load levels in ascending order and construct a discrete probability distribution function over the load levels.
4. For every system component, construct a discrete probability distribution over the capacity states that the component can assume.
5. Initialize sample index $i = 1$ and sample size $N = 1$.
6. For the sample index i, draw a system state x_i as follows:
   (a) Generate a random number and use it to draw a maintenance interval.
   (b) Generate a random number and use it to draw a load level within the maintenance interval drawn in step 6(a).
   (c) For every component that is *not* on maintenance in the interval drawn in step 6(a), draw a random number and use it to determine the capacity of the component.
   The combination of component capacities and load levels sampled in steps 6(b) and 6(c) describes the system state x_i. Note that this manner of sampling ensures that the probability of drawing x_i equals the actual probability of the physical system assuming the corresponding state.
7. Determine if the system experiences loss of load in state x_i. If so, update estimates of indices and test for convergence as described in Section 6.4.1. If the indices have converged to the specified tolerance, then end the simulation and report results. Otherwise increase the sample index i and sample size N and go to step 6.

This algorithm serves to convey an idea of the mechanics of the nonsequential simulation method. Here again, only one load curve has been assumed; the extension to different load curves at different buses is relatively straightforward. Section 6.4.1 outlines the approach for estimating indices from nonsequential simulation and testing for convergence, and this general approach can be used for estimating bulk power system reliability indices such as LOLP, LOLF, EUE and EPNS. In Section 11.5.2, the method of estimating LOLP, LOLF and EPNS are formalized.

In step 7, a testing of system performance is involved. Different methods for testing of system states are discussed in Section 11.6.

11.5.2 Estimation of Indices

The estimation of reliability indices from nonsequential simulation is based on the concepts described in Section 10.3.3, with two differences: (a) in Section 10.3.3 we were dealing with sets of states, whereas here we use individual states; (b) in nonsequential simulation all states are not evaluated, so we must apply statistical methods to estimate indices from the states that are sampled.

Theoretical Approach

For the purpose explaining how reliability indices are estimated, the system is initially assumed to be coherent, i.e., the failure or degradation of a component cannot improve system performance, and, likewise, the improvement or restoration of a component cannot deteriorate system performance. The means of enforcing this condition, as well as the consequences of relaxing this condition, will be discussed later.

As before (in Section 10.3.3), two more conditions will be imposed on the system:

(a) Every component can be represented by a two-state Markov model. We will show later that this approach can be extended to multistate components.
(b) The system can transit in one step from one state to another only by failure or restoration of a single component.

A transition that results from the repair of a failed component is said to be an *upward* transition; similarly, a transition that results from the failure of a functional component is said to be a *downward* transition.

We define the following abbreviations:

a-component: functional (available) component;
f-component: failed component;
A-state: functional (acceptable) state;
F-state: failed state.

Clearly, the probability of system failure (LOLP) is given by

$$P_F = \sum_{i=1}^{n_F} P\left\{x_i : x_i \in X_F\right\},$$ (11.12)

where X: set of all states;

X_F: set of system failure (loss of load) states; $X_F \subset X$;

n_F: number of failed states.

Each term in the summation described by (11.12) denotes an F-state, and each F-state describes a combination of a-components and f-components. It is possible for an F-state to transit to a higher state by an upward transition, and the resulting state may be an A-state or an F-state. If the resulting state is an A-state, then the transition is said to have crossed the boundary separating the F-states from the A-states. Now the frequency of encountering A-states is the sum of all the individual frequencies associated with those transitions of the F-states that cross the boundary.

Consider the sum described by (11.13), given below:

$$F^{+} = \sum_{i=1}^{n_F} \left[P\left\{x_i : x_i \in X_F\right\} \sum_{j \in Z_i} \mu_j \right],$$ (11.13)

where Z_i is the set of f-components in the ith state, and μ_j is the repair rate of jth component. F^{+} is the sum of the frequencies of all upward transitions from the failed states. Some of these transitions will cross the boundary, while others will not.

Now consider the sum described by (11.14), given below:

$$F^{-} = \sum_{i=1}^{n_F} \left[P\left\{x_i : x_i \in X_F\right\} \sum_{k \in \overline{Z}_i} \lambda_k \right],$$ (11.14)

where $\overline{Z}_i$ is the set of a-components in the ith state, and λ_k is the failure rate of the kth component. F^{-} represents the sum of the frequencies of all downward transitions of the failed states. However, because the system is coherent, *none* of these transitions will cross the boundary.

Next, consider any transition included in F^{-}, from state ℓ to state m, by failure of the rth component. Since all components are two-state Markov components, the following relationship holds:

$$P\left\{x_\ell\right\} \lambda_r = P\left\{x_m\right\} \mu_r.$$ (11.15)

Note that $P\{x_\ell\}\lambda_r$ is included in F^{-}, while $P\{x_m\}\mu_r$ is included in F^{+}. Observe also that this property extends to every transition included in F^{-}.

Now all those transitions included in F^+ that do not cross the boundary are balanced by corresponding transitions included in F^-; all the remaining transitions included in F^+ cross the boundary, and the sum of their frequencies gives F_S, the frequency of system success. At the same time, every transition included in F^- is balanced by an upward transition that fails to cross the boundary. Consequently, we have:

$$F_S = F^+ - F^-.$$

In the steady state, $F_F = F_S$, where F_F is the frequency of system failure. Therefore,

$$F_F = \sum_{i=1}^{n_F} \left[P\left\{x_i : x_i \in X_F\right\} \left(\sum_{j \in Z_i} \mu_j - \sum_{k \in \overline{Z}_i} \lambda_k \right) \right]. \tag{11.16}$$

Observe that we could have used a similar approach to compute F_F directly from the A-states, using

$$F_F = \sum_{i=1}^{n_S} \left[P\left\{x_i : x_i \in X_S\right\} \left(\sum_{k \in \overline{Z}_i} \lambda_k - \sum_{j \in Z_i} \mu_j \right) \right], \tag{11.17}$$

where X_S is the set of success states, and n_S is the number of success states.

Equation (11.16) is normally easier to use than (11.17), since for most systems $n_F < n_S$. Notice that (11.16) is in fact equivalent to (10.10). Notice also that (11.16) is contingent upon (11.15) being true. For two-state Markov components, frequency balance holds. In cases where the components assume multiple states, frequency balance does not always apply. However, in such cases, forced frequency balance can be imposed. This is achieved as described before in Section 10.3.3, by application of (10.1), repeated here for convenience. For every downward transition, say from system state ℓ to state m, by degradation of the rth component from component state s to state t, the actual downward transition rate λ_{st} is replaced by the fictitious transition rate λ'_{st}, given by $\lambda'_{st} = p_t \lambda_{ts}/p_s$, where p_t = probability of component r assuming state t, and λ_{ts} = actual upward transition rate of component r from state t to state s.

If we now apply the same arguments as before and use fictitious downward transition rates from the F-states to balance all those upward transitions that do not cross the boundary, we then have the following equivalent for (11.16), for the case of multi-state components:

$$F_F = \sum_{i=1}^{n_F} \left[P\left\{x_i : x_i \in X_F\right\} \sum_{j=1}^{n_C} \left(\lambda_{ji}^+ - \lambda_{ji}^- \right) \right], \tag{11.18}$$

where n_C: number of components in the system;

  λ_{ji}^+: transition rate of component j from its state in system state i to higher-capacity states;

  λ_{ji}^-: frequency-balanced transition rate of component j from its state in system state i to lower-capacity states.

The transition rates λ_{ji}^+ and λ_{ji}^- are given by:

$$\lambda_{ji}^+ = \sum_{s=r+1}^{n_j} \lambda_{rs} \quad \text{and} \quad \lambda_{ji}^- = \sum_{t=1}^{r-1} \lambda_{tr}',$$

where r is the state of component j in system state i, n_j is the number of states component j can assume and the λ' terms denote fictitious transition rates calculated using forced frequency balance as explained above.

It is possible to construct the frequency distributions of the components in such a manner that forced frequency balance applies, without having to individually compute all the fictitious transition rates. This will be described below when the practical implementation is explained. This approach will enable application of (11.18) in a simple and direct manner.

Equations (11.12) and (11.18) provide the exact probability and frequency of failure when all the failure states are known. We shall now describe how to adapt (11.12) and (11.18) to estimate LOLP and LOLF from states sampled during nonsequential Monte Carlo simulation. We have described in Section 11.5.1 the manner in which a random state $\{x_i \in X\}$ is sampled. Based on the sampled x_i, the values of the random variables π_i and ϕ_i are determined as follows:

$$\pi_i = \begin{cases} 1 & \text{if } x_i \in X_F \\ 0 & \text{otherwise} \end{cases} \tag{11.19}$$

$$\phi_i = \begin{cases} \sum_{j=1}^{n_C} \left(\lambda_{ji}^+ - \lambda_{ji}^- \right) & \text{if } x_i \in X_F \\ 0 & \text{otherwise} \end{cases}. \tag{11.20}$$

With each drawing, and subsequent increase in sample size N, the values of the random variables Π_N and Φ_N are determined as follows:

$$\Pi_N = \frac{1}{N} \sum_{i=1}^{N} \pi_i \tag{11.21}$$

$$\Phi_N = \frac{1}{N} \sum_{i=1}^{N} \phi_i. \tag{11.22}$$

Upon convergence, Π_N and Φ_N provide the estimates of P_F and F_F, respectively; these failure indices, which are basically the system LOLP and LOLF, are given by:

$$P_F = E\left[\Pi_N\right] \tag{11.23}$$

$$F_F = E\left[\Phi_N\right], \tag{11.24}$$

where $E[\cdot]$ is the expectation operator. The estimates Π_N and Φ_N are said to have converged to the statistics P_F and F_F when the sample size N is large enough that the coefficients of variation of the estimates are arbitrarily small, i.e., when the following relationships hold:

$$\frac{\sqrt{Var\left(\Pi_N\right)}}{E\left[\Pi_N\right]} \leq \delta \tag{11.25}$$

$$\frac{\sqrt{Var\left(\Phi_N\right)}}{E\left[\Phi_N\right]} \leq \varepsilon. \tag{11.26}$$

where $Var(\cdot)$ is the variance function, and δ and ε are prespecified tolerances.

Practical Implementation

We have stated earlier, in Section 11.5.1, that the generation, transmission and load are represented by discrete distributions of probability and frequency corresponding to the various capacity levels. Based on the capacity states and interstate transition rates of generating units available at a given bus, discrete probability and frequency distributions are constructed, at every bus, using the *Unit Addition Algorithm* described in Chapter 8. Since these distributions are cumulative, it can be stated that:

$$P_j(n) = P_j(n-1) + p_j(n) \tag{11.27}$$

$$F_j(n) = F_j(n-1) + p_j(n)\left(\lambda_{jn}^+ - \lambda_{jn}^-\right), \tag{11.28}$$

where $P_j(n)$: cumulative probability of the nth state;
  $F_j(n)$: cumulative frequency of the nth state;
  $p_j(n)$: exact state probability of the nth state;

and the subscript j is the component index.

  (A less general form of (11.28), applicable to two-state generating units, had been derived in [3]. The generalized, multistate form given by (11.28) was developed later in [7].)

An array of "incremental transition rates," L_j, is constructed from the cumulative frequencies, using equations (11.27) and (11.28):

$$L_j(n) = \frac{F_j(n) - F_j(n-1)}{P_j(n) - P_j(n-1)} \tag{11.29}$$

Therefore in practical implementation, 11.20 is applied in the following form:

$$\phi_i = \begin{cases} \sum_{j=1}^{n_C} L_j(n) & \text{if } x_i \in X_F \\ 0 & \text{otherwise} \end{cases}. \tag{11.30}$$

The distributions P, F and L need to be constructed only once, at the beginning of the program. Thereafter the simulation algorithm is executed, with the estimates being updated as described above.

Other Indices
The LOLP and LOLF indices are estimated as P_F and F_F, given by (11.23) and (11.24). Other indices, such as MDT and EPNS, can be estimated as described below.

Mean down time (MDT)
The MDT can simply be estimated from:

$$\text{MDT} = \frac{\text{LOLP}}{\text{LOLF}}. \tag{11.31}$$

Expected power not served (EPNS)
The EPNS was mathematically defined in (11.9). Using nonsequential simulation, this index is estimated as follows. Based on the sampled x_i, the value of the random variable ψ_i is determined as:

$$\psi_i = \begin{cases} C\{x_i\} & \text{if } x_i \in X_F \\ 0 & \text{otherwise} \end{cases}, \tag{11.32}$$

where $C\{x_i\}$ is the system load curtailment in the ith sampled state. Then, EPNS is estimated from:

$$\text{EPNS} = E\left[\Psi_N\right]; \quad \Psi_N = \frac{1}{N}\sum_{i=1}^{N} \psi_i, \tag{11.33}$$

where Ψ_N is the estimator of EPNS after N samples.

11.6 Testing of States

In all types of simulation, system states need to be tested to determine if there is loss of load, and if so, how much. In most cases, the testing is based on the assumption that for a given system contingency, the operator will attempt a feasible, i.e., security constrained, dispatch that avoids or minimizes the load shedding without violating system constraints such as equipment loading limits and bus voltage limits. This is customarily formulated as a constrained optimization problem, where the objective function is the load curtailment and the problem consists of minimizing the total system curtailment subject to (a) power flow constraints and (b) security constraints. Certain operating policies, such as firm contracts, emergency assistance policies, etc., are also modeled and included in the constraints. The elements of security constrained dispatch have been summarized in Section 9.4.

The solution to the security constrained dispatch problem as described in Section 9.4.6 provides the minimum load curtailment P_C for the given contingency. If this value of P_C is zero, then the contingency x_i represents an acceptable state; otherwise it is a loss of load state, and the value of P_C provides the $C\{x_i\}$ for the state x_i. This $C\{x_i\}$ is used in (11.10) or (11.33) to estimate the system EPNS.

For power systems of realistic size, the complete, nonlinear optimization problem can often get very large and time consuming. This is an important consideration, since the problem needs to be solved for every state tested, and the number of states that must be tested before the simulation converges is quite large. In fact, the better-designed and more reliable the system, the larger the number of states that must be tested before the indices converge. Consequently, attempts have been made to reduce the computational effort expended on testing states.

Simplified alternatives to the complete nonlinear optimization fall into two broad classes: linear approximations and heuristic methods. The former consists of simplifying the nonlinear power flow equations into linearized flow approximations, such as (a) the transportation model (also known as capacity flow model, network flow model and contract path model), where only the equipment capacity limits are considered, and the dependence of the flows on the line and transformer impedances are ignored; and (b) the DC flow model, where a linear dependence is assumed between the real power transfers and the bus voltage angles. These methods have been outlined in Section 9.4 and described in some detail in [47]. In linear approximation methods, the reactive power flows and voltage constraints are ignored. The second class, i.e., heuristic methods, has several variations. In some instances, simple, rule-based dispatches have been implemented; in others, neural networks have been trained and used for testing of states. An example of using neural net-based methods for characterizing states during Monte Carlo simulation can be found in [60].

11.7 Acceleration of Convergence

Inevitably, attempts have been made to reduce the number of samples required to converge to a desired accuracy. The various methods that have been successfully applied to the problem of power system reliability assessment using Monte Carlo simulation are: (a) control variates [61], (b) antithetic variables [61], stratified sampling [62], importance sampling [63], cross-entropy [17, 64], etc. These methods are based on variance reduction techniques and have been briefly described in Section 6.5. In the next section we provide a description of the method of state space pruning [9], which is a Monte Carlo method with highly accelerated convergence.

11.8 State Space Pruning: Concept and Method

State space pruning is a method of accelerating the nonsequential Monte Carlo simulation approach outlined in Section 11.5. It consists of performing Monte Carlo simulation selectively on those regions of the state space where loss of load states are more likely to occur. These regions are isolated by identifying and removing coherent acceptable subspaces. The acceptable sets may be identified by any suitable means: while the method described in Section 10.3, using the lower bound of acceptability, is the most aggressive, heuristic methods have also been used for algorithmic simplicity [65]; it may also be expedient to simply use random sampling only to identify acceptable states and prune out all states of higher capacity.

11.8.1 The Concept of Pruning and Simulation

For a system of realistic size, the state space would be very large, and if the system is highly reliable, as most systems are designed to be, the number of states sampled during Monte Carlo simulation would have to be quite large before the indices converge to reasonably small tolerances. Depending on the complexity of the system model, and the reliability of the system, the process of testing a large number of samples may turn out to be extremely time consuming. In the method of pruning and simulation, arbitrarily large parts of the state space are identified where failure states do not occur, and these sets of acceptable states (A-sets) are pruned off prior to performing simulation on the remaining state space. This increases the likelihood of a sampled state being an F-state, since the proportional probability of every state with respect to the sample space is increased. Naturally, the simulation converges much faster.

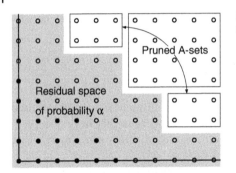

Figure 11.1 Residual state space. Solid circles indicate failed states.

This can be understood as follows. Consider a hypothetical two-dimensional discrete state space over which it is desired to perform Monte Carlo simulation to obtain reliability indices. Using some suitable device, an arbitrary number of coherent acceptable sets are identified and removed. This "pruning" results in a reduced state space, wherein the loss of load states are left undisturbed, as in the original state space, but the likelihood of encountering these loss of load states in the course of Monte Carlo sampling over the residual state space is higher than the likelihood of the same in the original state space. This is depicted in Figure 11.1. It is reasonable to assert that simulation over the residual state space would require a smaller sample size to meet the same convergence criterion.

This assertion is justified as follows. Consider calculating the *system Loss of Load Probability* (LOLP), **p**;

Let p be the estimate of **p**;

Since failure states are binomially distributed (a sampled state can be a failure state or a success state), the distribution has mean p and variance $p(1 - p)$;

Then the coefficient of variation of p is

$$\eta = \frac{1}{p}\sqrt{\frac{p(1-p)}{N}}, \tag{11.34}$$

where N is the number of sampled states.

If sampling is performed over the residual state space of probability α, then the estimate of the *conditional* LOLP is

$$p' = \frac{p}{\alpha}.$$

Then the coefficient of variation of p' is

$$\eta' = \frac{1}{p'}\sqrt{\frac{p'(1-p')}{N'}} = \frac{\alpha}{p}\sqrt{\frac{\frac{p}{\alpha}\left(1 - \frac{p}{\alpha}\right)}{N'}}, \tag{11.35}$$

where N' is the number of states sampled from the residual state space.

Now if both the estimates p and p' are required to converge to the same tolerance, then

$$\frac{N'}{N} = \alpha \frac{1 - \frac{p}{\alpha}}{1 - p}. \tag{11.36}$$

Note that (11.36) has been obtained by equating η and η', i.e., the coefficients of variation of p and p', which are actually estimates of different quantities. However, η and η' are the coefficients of variation within their respective state spaces over which sampling was performed, and it is therefore reasonable to equate them.

Note also how (11.36) indicates an approximate equality between the fraction N'/N and the residual probability α, since p is small. In other words, the reduction in sample size is almost proportional to the extent of pruning the state space.

Pruning is performed by identifying an acceptable state (partition vector) and discarding this state and all higher states. This may be performed as many times as desired or deemed suitable. The manner in which the partition vector is determined depends on how the system is modeled, and on how the network flows are represented. In composite system reliability modeling, the network is modeled in the form of nodes interconnected by arcs, where the nodes and arcs represent the buses and transmission lines, respectively. The generation and load at each node and the transmission capacity of each arc is respectively modeled as a set of discrete capacity levels, each level associated with a probability and frequency. These probability and frequency distributions are constructed as cumulative distributions, so that frequency balance is enforced.

State Space and Partition Vectors
For a given load scenario, the available bus generations and the transmission line capacities will determine whether the bus loads will be satisfied. We can therefore define the state space as the set of all possible combinations of generation levels and transmission line capacities. The capacities of the generation and transmission components in each state comprise the elements of the vector describing that state. The treatment of temporal load variations and planned outages of generators can be accommodated using load clustering, as described in Section 10.3; however, this is summarized here for convenience. The approach involves clustering the multinode hourly loads into a relatively small number (10–20) of multinode load vectors, each with an associated probability and frequency. Then a reference load state is defined, using which the generation model is modified for every load cluster, and these modified models are interleaved to construct an integrated generation model. Similar modification and subsequent integration is used to accommodate planned outages. The resulting generation model needs to use only one load state, which is the

reference state. The method is described in detail in Section 10.3.2. Since clustering finally results in a model that uses a single load level, the pruning and simulation method will be described here using a single load level. The idea can be easily extended to include multiple load levels and planned maintenance. In general, therefore, a system of N_n nodes and N_a transmission lines (arcs) will have a discrete state space of dimension $(N_n + N_a)$, with each axis consisting of bus generation or transmission line capacity levels, zero levels included.

As mentioned earlier, pruning can be performed by identifying and removing several A-sets. A-sets may be identified most aggressively using the same approach as in state space decomposition outlined in Section 10.3. The original state space is first treated as an unclassified set (U-set). Based on the maximum capacity levels available in this U-set, the system load curtailment is minimized; then the combination of the lowest-capacity states that yield zero curtailment constitutes a partition vector to which we have previously referred as the u-vector. The u-vector has the property that all states between and including the u-vector and the upper boundary of the U-set will be acceptable states and will constitute an A-set. Using the u-vector, the original U-set is now decomposed into an A-set, and $N_n + N_a$ disjoint U-sets, some of which may be empty. (Note that in using the decomposition approach for pruning, v-vectors and L-sets are not identified, thereby avoiding the bulk of the effort required by the method of state space decomposition.) The A-set is discarded, and the remaining U-sets are submitted to simulation. If more A-sets are desired to be removed, more U-sets are decomposed. As the pruning progresses, the size of the A-sets generated diminishes, but the effort expended on determination of a u-vector (which requires solution of a constrained optimization problem) remains the same. So it is prudent to stop pruning after a point, and perform sampling over the residual state space.

There are alternative means of finding u-vectors. Heuristic methods have also been used for algorithmic simplicity. Reference [65] describes a method where the generation available at the nodes are proportionately reduced so that their sum equals the total load; if the resulting nodal generations satisfy power flow constraints, then they comprise the u-vector for the given U-set. If power flow constraints are not satisfied, the heuristic fails and the u-vector has to be determined using the constrained optimization method. (Note that if generation does not exceed load, then the U-set contains no u-vector.) It may also be expedient to simply use random sampling and test the sampled state for acceptability; if there is no loss of load, the sampled state can be treated as a u-vector. This approach is the least aggressive (and therefore least efficient in terms of the size of A-sets pruned), but involves the least effort.

From the undecomposed U-sets (comprising the residual space) a U-set is randomly selected in such a manner that the likelihood of selecting that state equals the ratio of the probability of the U-set to that of the residual space. Within the selected U-set, proportional sampling is used to select a generation level at every node, and a transmission capacity for every arc. If multiple load levels are used, a load level is also sampled. The generation-transmission-load scenario thus selected constitutes the sampled state, which is tested for acceptability, and (11.19)–(11.22) are used to update the estimates of the system indices. Similar criteria are used to estimate the failure indices at the buses. It is important to state at this point that (11.19)–(11.22) estimate indices with respect to the sample space, so these estimates must be divided by the residual space probability α to get the correct estimates of P_F and F_F.

11.8.2 Dealing with Noncoherence

In pruning, as in decomposition, the state space needs to be coherent, i.e., an upward transition from an A-state should result in an A-state, and a downward transition from an F-state should result in an F-state. If this condition is not satisfied, then (a) the arguments presented in Section 11.5.2 for estimating the frequency indices cease to be valid, and (b) the process of pruning, which depends on all states higher than the u-vector being A-states, fails.

Power flow models other than the capacity flow model can result in noncoherence. This is because, when using other power flow models, transmission capacity changes are accompanied by changes in line impedances, causing the power flows to be redistributed. So if a system is in an A-state, the redistribution of flows resulting from the restoration of a failed transmission line may cause loss of load in some part of the system; similarly, a line failure may result in a transition from an F-state to an A-state. We deal with this problem as follows. During the pruning phase, the decomposition is performed over the generation levels, holding the transmission levels at the highest-capacity states. In other words, every time the u-vector is determined, the components of the u-vector corresponding to the transmission lines are set at the maximum capacity levels. Since the transmission system is generally far more reliable than the generation system, fixing the transmission levels does not significantly reduce the effectiveness of the pruning. This takes care of problem (b) mentioned above.

To deal with problem (a), we construct a premise, based on the following propositions:

Proposition 1: Dispatcher action precludes the occurrence of those events whereby the restoration of a failed component results in deterioration of system performance.

Proposition 2: In a power system rendered noncoherent solely by inclusion of DC or AC flow constraints, the aggregated probability of states contributing to noncoherence is negligibly small.

Proposition 1 simply means that a dispatcher can choose not to bring a repaired transmission line back into service, if such restoration is deemed detrimental to system performance. We claim that proposition 1 is justifiable from an engineering standpoint, and as such does not represent an approximation or inaccuracy. However, it excludes those events whereby the degradation or failure of a component improves system performance, since these are random events beyond the dispatcher's control. Ignoring these transitions constitutes an approximation which, we claim via proposition 2, may be acceptable. The results reported in [9, 66] support this assertion. Besides, when pruning is performed, these states always reside in the residual space, so if the indices are estimated only from sampled F-states, these states, if sampled, are appropriately identified as A-states, and contribute to no error at all.

Nevertheless, the above premise represents a mathematical imperfection, and its acceptance or otherwise is essentially a matter of engineering judgement. If, however, we choose to accept it, then problem (a) is solved, because the method described in Section 11.5.2 computes the frequency indices from sampled F-states, and pruning the state space leaves *all* F-states for simulation. We emphasize the word *all*, since we have chosen to accept proposition 1.

Treatment of Multistate Loads

The modeling of loads is a little more complex, unless only one load state is used. If a multistate load model is used, then the estimation of frequency indices will be incorrect unless all nodal loads are perfectly correlated, i.e., unless they all increase and decrease together; otherwise the residual state space will be noncoherent. In real life, temporal load profiles are bound to be diverse across geographically extensive systems. However, in composite system analysis we normally use a detailed representation of only a part of a system, and if the load diversity is not too large, then the LOLF can be computed, albeit with reduced accuracy.

In constructing the P, F and L distributions (see Section 11.5.2) for a multinode, multistate load model, the clustering approach is used (see Section 9.3.3). There are two more points to note in the context of load modeling. One is that in order that system coherence is preserved, load states are sequenced in the reverse order, i.e., a higher load level is regarded as a lower load state, since a decrease in system load improves the system performance from a reliability perspective. The other point to note is that since a multistate load model is constructed in the form of multinode vectors, the entire load model is associated with only one set of P, F and L distributions. Consequently, in constructing these distributions, the number of components is $n_C = N_n + N_a + 1$ for a multistate load model, where the additional dimension, corresponding to the load, has as many states as there are load clusters.

11.9 Intelligent Search Techniques

While Monte Carlo simulation consists of random searches in the system state space, some recent efforts have investigated the use of nonrandom or intelligent searches, based on heuristic optimization methods. In this book we briefly describe methods using genetic algorithms, particle swarm optimization and neural networks. A comparison of several techniques is provided in [67].

11.9.1 Genetic Algorithms

A genetic algorithm (GA) is a simulation of evolution where the rule of survival of the fittest is applied to a population of individuals. In the basic genetic algorithm an initial population is randomly created from a certain number of chromosomes. In power systems [68, 69], each chromosome represents a system state. Each chromosome consists of binary number genes. Each gene represents a system component. If any gene takes a *zero* value this means that the component it represents is in the *down* state and if it takes a *one* value that means its component is in the *up* state.

All of the individuals are evaluated using a certain fitness function. A new population is selected from the old population based on the fitness of the individuals. Some genetic operators, e.g., mutation and crossover, are applied to members of the population to create new individuals. Newly selected and created individuals are again evaluated to produce a new generation and so on until the termination criterion has been satisfied.

The suitable choice for the fitness function can add the required intelligence to GA state sampling. Since we are searching for high-probability failure states, the fitness function should be so constructed as to pick and propagate states with high failure probability. One possible choice of the fitness function is given below:

$$fit_j = \begin{cases} SP_j & \text{if new chromosome } j \text{ represents a failure state} \\ SP_j \cdot \beta & \text{if old chromosome } j \text{ represents a failure state} \\ SP_j \cdot \alpha & \text{if new or old chromosome } j \text{ represents a success state} \\ SP_j & \text{if chromosome probability is below threshold value} \end{cases}, \quad (11.37)$$

where SP_j is the state probability, β is a small number in the range of 0.1 to 0.0001, and α is a very small number, e.g., 10^{-20}. In this manner the fitness value of an old failure chromosome is reduced to enable GA to search for more failure states and prevent failure states with higher probability from dominating other failure states. The fitness function is scaled to enhance the performance of the GA search process.

After calculating the fitness value of all chromosomes in the current population, GA operators are applied to evolve a new generation. These operators are selection schema, crossover and mutation. New GA generations are produced until a stopping criterion is reached. The main role of GA is to truncate the huge state space by tracing states that contribute most to system failure. After

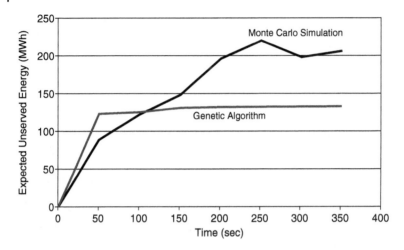

Figure 11.2 Comparison of convergence characteristics of Monte Carlo simulation and genetic algorithm.

the search process stops, data saved in the state array is used to calculate the annualized adequacy indices for the whole system and for each load bus.

One should note an important difference between Monte Carlo simulation (MCS) and GA. MCS relies on a random search over the state space, while GA uses an "intelligent" search. The "random" property of MCS causes the estimator to sometimes overestimate and sometimes underestimate the corresponding index, so that the estimator approaches the index from both sides; GA, on the other hand, accumulates the estimate as it proceeds, based on the contributions of the states visited, so that as it approaches the index, it is always less than the actual value of the index. This is illustrated in Figure 11.2, which shows the results obtained in [69].

11.9.2 Particle Swarm Optimization

Particle swarm optimization (PSO) [70] was developed as a result of simulating social behavior of animals displaying group intelligence, such as that of a swarm of birds in search of food. Classically, the simulation is performed on a two-dimensional grid (representing an area of land, where food is supposed to be located). The birds set out in search of food, the location of which is unknown to them. They follow certain dynamics and concur upon the position where the food is located. In this paradigm, the particles (birds) traverse (fly over) the state space (an area of land) in search of the optimum (food). Each point in the space has a fitness value (a sense of distance from the location of the food). The motion of the particles is governed by three aspects: one, inertia, i.e., they continue to move in the direction in which they had been moving; two, local optima: they tend to move towards the position that gives their personal best fitness value; and three, the group optima: they tend to move towards the position that gives the best fitness within the same group. The interaction of these three factors gives the particles new positions and in due course of time they settle upon the position that gives the best fitness value (zero distance from the food). In the process they visit other positions as well. In reliability analysis, these visits are used to collect reliability data.

In adapting the method to power system reliability analysis [71], the PSO technique is used as a search tool instead of an optimization method. The state space is constructed in a manner similar to that used in nonsequential simulation. A swarm of particles starts out from randomly initialized positions in the state space and preferentially seeks out loss of load states. As loss of load states are visited, their contributions to reliability indices are accumulated. In its simplest formulation, the algorithm is a fitness function similar to that used by the GA method described above, and the swarm seeks out the significantly contributing loss of load states. A multi-objective formulation, with one fitness function based on the state probability and another based on the system load curtailment, was found to exhibit better swarm dynamics. The reason is that high-probability failure states typically produce lower curtailment, and vice-versa. So the use of multiple and conflicting objectives ensure that the particles would never converge upon any one part of the state space and thereby end the search; rather, the conflicting objectives force the particles to continue their selective search for failure states until the estimated indices converge.

Since this method, too, estimates reliability indices by accumulating the contributions of the failure states visited, the convergence characteristic of this method is also one-sided, like that of the GA-based method.

11.9.3 Neural Networks

From the steps of the straight Monte Carlo simulation, we can make two observations: 1) for each sampled state, a flow calculation has to be performed to determine its load-loss status; 2) because of the random sampling, many similar states are sampled in the simulation and their characteristics determined repeatedly. Therefore, the straight Monte Carlo simulation is very time consuming. Three methods have been described in [72–74].

One method is to more efficiently determine the load loss characteristic of the sampled state. In this method, the *Self-Organizing Map* (SOM) is trained to recognize the loss-of-load states. Once this training is complete, the SOM is used along with the Monte Carlo simulation to estimate the reliability of the system. This method overcomes the first disadvantage of the straight Monte Carlo simulation. Another method that has been used to identify system states in flexible manufacturing systems is based on *Group Method of Data*

Handling (GMDH). This method can also be easily used for power system applications [75].

The second method proposes to cluster the sampled states before determining their load loss characteristics. In this method, Monte Carlo simulation is used first to accumulate states, then SOM is used to cluster these states, and flow calculation is used for analysis of clustered states. This method overcomes the second disadvantage of the straight Monte Carlo simulation.

11.10   Conclusion

It is appropriate to make the following remarks before concluding this chapter.

A significant majority of reliability analysis programs used in the community—both commercial software packages used in the industry and research-grade programs used by the research community—utilize one of the Monte Carlo algorithms described in this chapter or variations thereof. Further, most probabilistic production-costing programs used in the community also utilize similar algorithms, with the difference that the objective function of maximizing reliability or minimizing curtailment is replaced by that of maximizing the social welfare function (in simulating competitive markets with demand elasticity) or minimizing the production cost (in simulating regulated regimes or markets without demand elasticity).

In both reliability and production simulation, the use of optimization functions affords the ability to determine not only system-wide indices as described in this chapter, but also locational indices, such as nodal or zonal reliability indices or marginal prices. Locational reliability indices can be obtained by accumulating locational information (such as curtailment) over the samples or events. Locational marginal prices and other sensitivity indices [76] can be obtained from the values of the Lagrangian (or dual) variables obtained at the optimal solutions over the course of the simulation.

Finally, a cautionary statement about Monte Carlo algorithms. It is important to remember that each Monte Carlo realization (or replication) represents only one sample path, and that in instances where a realization is constructed over a planning horizon over which resources and demands are known or projected, it is necessary to simulate several (sometimes hundreds or even thousands) realizations before the estimates converge to acceptable tolerances. This cautionary statement applies to those commercial programs in which the test for convergence is not effectively implemented or explained; often, the software provides the user with the option of executing the simulation over only one sample path, without warning the user that any statistical inference drawn from one sample path is almost guaranteed to be grossly inaccurate. It is a good idea for the educated user to bear this in mind and avoid the temptation of executing only one replication of what is often a very time-consuming simulation.

12

Power System Reliability Considerations in Energy Planning

12.1 Introduction

We have discussed the reliability criteria and the methods of reliability evaluation in previous chapters. Ultimately reliability analysis provides an input to the decision-making process. As an example, here we discuss how to incorporate reliability considerations into a power system expansion planning problem. Power system reliability indices can be broadly categorized as probabilistic and deterministic. By their nature the probabilistic techniques are more suitable for applications like trade-offs and resource optimization, as they respond effectively to the changes in the unit mix, location and load shape. We propose a stochastic programming framework to effectively incorporate random uncertainties in generation, transmission line capacity and system load for the expansion problem. Favorable system reliability and cost trade-off is achieved by the optimal solution. The problem is formulated as a two-stage recourse model where random uncertainties in area generation, transmission lines and area loads are considered. A power system network is modeled using DC power flow analysis. The reliability index used in this problem is the expected cost of load loss as it incorporates duration and magnitude of load loss. Due to the exponentially large number of system states (scenarios) in large power systems, we apply the sample-average approximation (SAA) concept to make the problem computationally tractable. The method is implemented on the 24-bus IEEE reliability test system.

Energy planning is the development of policy to ensure medium- to long-term energy supply and delivery to end consumers. The generation expansion problem addresses critical issues of optimal location for new generation resources. This information can be used by entities like Independent System Operators (ISO) to generate price signals or other incentives for materializing such resources.

Electric Power Grid Reliability Evaluation: Models and Methods, First Edition. Chanan Singh, Panida Jirutitijaroen, and Joydeep Mitra.

Reliability evaluation of power systems can be considered and characterized in two aspects: deterministic and probabilistic. Deterministic indices are rules of thumb, such as reserve as a percentage of the peak load or equal to the capacity of the largest unit. While the deterministic indices are easy to assess and implement, they tend to provide an unreasoned margin of safety and are not suitable to trade-off reliability and cost. Probabilistic indices, such as loss of load probability or expected energy not supplied, on the other hand, are much more complicated to incorporate in the problem, but they can represent the quality of potential solutions in a more comprehensive and mathematically based manner. The integration of probabilistic reliability assessment is thus the focus of this chapter.

Several optimization techniques have been proposed for the generation expansion problem [77]. We mainly consider those with explicit formulation of uncertainties. Among them, [78, 79] propose stochastic programming framework for the planning problem and formulate it as a two-stage recourse model. The first-stage decision on expansion policy is completed before the random uncertainties in generation, transmission line capacity and load are realized. After the random realization, i.e., the generating capacity, load demand and transmission capacity are known, the second-stage decision is to determine the generating capacity from each bus to minimize operation cost with the assumption that the load is satisfied at all times. The problem is solved with a large-scale deterministic equivalent problem.

In this chapter, we modify the formulation and incorporate reliability considerations in the planning problem [80, 81]. The reliability index used is expected cost of load loss, which is computed in the second-stage problem. The term load loss is used to indicate the load that could not be served because of generation or transmission deficiency. The overall objective is then to minimize expansion cost in the first-stage and at the same time to minimize operation cost and expected cost of load loss in the second-stage. Generating unit availability of additional units in terms of their forced outage rates and derated forced outage rates can also be included in this formulation.

The deterministic equivalent problem under a stochastic programming framework becomes a large-scale linear programming problem with special structure. The L-shaped algorithm is a standard algorithm for solving large-scale stochastic programming problems. Interested readers may refer to [82, 83] for details of the two-stage recourse model formulation and available solution algorithms. The algorithm considers the entire probability space which, for a very large system, may be impractical and even impossible to enumerate and evaluate. Direct application of the L-shaped algorithm cannot thus be achieved in a computationally effective manner.

A technique called sample-average approximation is used to overcome this problem of dimensionality. The sampling technique is employed to reduce the number of system states. The objective function of the second-stage, called sample-average approximation (SAA) of the actual expected value, is defined by these samples. This approximation makes it possible to solve the problem with the deterministic equivalent model. The objective function values from the SAA problems are in fact estimates of the actual optimal objective values. These estimates yield upper bounds and lower bounds of the optimal objective value (actual expected value). Upper bounds and lower bounds of the optimal objective values can then be used to analyze the quality of the approximate solutions.

We also analyze in Section 12.4 the performance of the optimal solution to reliability distribution of each node to examine if the system-wide reliability maximization can lead to a fair reliability enhancement for all customers [84]. When each customer is charged equally the expansion cost, each may not receive the same reliability level. Since the formulation considers reliability in terms of the overall system, the optimal solution yields system reliability maximization, which may not guarantee fair improvement of reliability for customers at each bus.

The chapter is organized as follows. Detailed problem formulation is first introduced. The sample-average approximation technique is described next. Lower and upper bound estimation of the actual objective value and the approximate solution are presented and discussed. The method is implemented on a 24-bus IEEE-Reliability test system [35]. A comparative study of system-wide reliability maximization is examined. Concluding remarks are given in the last section.

12.2 Problem Formulation

The objective of energy expansion problem is to minimize the expansion cost while maximizing system reliability under uncertainty in generation, load and transmission lines. When reliability is one of the constituents, a reliability model needs to be incorporated into the problem formulation. This underlying reliability evaluation model requires a flow model for evaluating system states regarding their loss of load status. A commonly accepted approach in composite reliability evaluation is to use a DC power flow model. The capacity of every element in the network is represented by a random variable with its discrete probability distribution. Using system expected cost of load loss as a reliability index, the problem is formulated as a two-stage recourse model.

A Generation Expansion Planning Problem with Two-Stage Recourse Model
The first-stage decision variables are the number of generators, x_i , to be installed at node i, while the second-stage decision variables are the power at nodes and flows in the network, i.e., power generation, power flow in transmission lines and load curtailment in each system state. The first-stage decision

variables in the expansion policy are determined before the realization of randomness in the problem, while the second-stage decision variables are evaluated after the random uncertainties are realized. The objective of the problem is given by (12.1) to minimize both the expansion and operation cost:

$$\min \sum_{i \in I} c_i x_i + E_{\tilde{\omega}} \{f(x, \tilde{\omega})\} \tag{12.1}$$

$$s.t. \sum_{i \in I} c_i x_i \leq B \tag{12.2}$$

$$x_i \geq 0, \text{integer}, \tag{12.3}$$

where c_i is cost of additional generators at bus i and (12.2) represents a budget constraint of B. Constraint (12.3) is a requirement for the number of additional generators to be an integer.

The function, $E_{\tilde{\omega}} \{f(x, \tilde{\omega})\}$, in (12.1) is the expected value of the second-stage objective function to minimize operation cost and loss of load cost under a realization ω of state space Ω. The second-stage problem is to schedule the generating capacity in order to minimize operation cost and the reliability cost incurred from the load curtailment in each state. Power system network constraints are formulated using DC flow model. Second-stage decision variables are: generation at bus i in state ω, $y_{gi}(\omega)$, load curtailment at bus i in state ω, $y_{li}(\omega)$ and voltage angle at bus i in state ω, $\theta_i(\omega)$:

$$f(x, \omega) = \min \sum_{i \in I} \{c_{li}(\omega) y_{li}(\omega) + c_{oi}(\omega) y_{gi}(\omega)\} \tag{12.4}$$

$$s.t. y_{gi}(\omega) \leq g_i(\omega) + A_i x_i, \forall i \in I \tag{12.5}$$

$$b_{ij}(\theta_i(\omega) - \theta_j(\omega)) \leq t_{ij}(\omega), \forall i, j \in I, i \neq j \tag{12.6}$$

$$y_{li}(\omega) \leq l_i(\omega), \forall i \in I \tag{12.7}$$

$$\sum_{j \in I} B_{ij} \theta_j(\omega) + y_{gi}(\omega) + y_{li}(\omega) = l_i(\omega), \forall i \in I \tag{12.8}$$

$$y_{gi}(\omega), y_{li}(\omega) \geq 0, \forall i \in I, \theta_i(\omega) \text{unrestrict}, \tag{12.9}$$

where $c_{li}(\omega)$ and $c_{oi}(\omega)$ are cost of load loss and cost of operation at bus i in state ω in dollar per MW, A_i is an additional generation capacity at bus i in MW. Parameters $g_i(\omega)$, $t_{ij}(\omega)$ and $l_i(\omega)$ are generation capacity at bus i in MW, tie-line capacity between buses i and j in MW and load at bus i in MW in state ω. $\vec{B} = [B_{ij}]$ is an augmented node susceptance matrix and b_{ij} is tie-line susceptance between bus i and bus j. It should be noted that the cost of the load loss coefficient depends on system states. The calculation of this coefficient is performed separately and is shown next.

Constraints (12.5), (12.6) and (12.7) are maximum capacity flows in the network under uncertainty in generation, tie-line and load, respectively. Constraint (12.8) constitutes conservation of flow in the network, and (12.9) presents variable restrictions in the model. Note that the decision on the expansion policy is done before the realization ω of Ω.

The failure probability of additional generators can be taken into account by using their effective capacities or by explicitly incorporating the unit availability of additional units in terms of their forced outage rates and derated forced outage rates in the formulation. The first-stage problem is slightly modified in as follows:

$$\min \sum_{i\in I} \sum_{q\in Q_i} c_{iq}x_{iq} + E_{\tilde{\omega}}\{f(x, \tilde{\omega})\} \tag{12.10}$$

$$s.t. \sum_{i\in I} \sum_{q\in Q_i} c_{iq}x_{iq} \leq B \tag{12.11}$$

$$x_{iq}, \text{binary}, \tag{12.12}$$

where c_{iq} is the cost of additional generators q at bus i, x_{iq} is equal to 1 if generating unit q is installed in area i and 0 otherwise, and Q_i is the total number of additional generators at bus i. Constraint (12.5) in the second-stage problem only needs to be modified as follows:

$$y_{gi}(\omega) \leq g_i(\omega) + \sum_{q\in Q_i} A_{iq}(\omega)x_{iq}, \forall i \in I, \tag{12.13}$$

where A_{iq} is additional generating capacity of unit q at bus i in state ω in MW. It should be noted that the number of system states without unit availability of additional generators is much less than that with unit availability consideration, as we shall see from Section 12.4.

Loss of Load Cost (LOLC) Coefficient Calculation

Loss of load cost depends on interruption duration as well as the type of interrupted load. The most common approach to represent power interruption cost is through the customer damage function (CDF) [7]. This function relates different types of load and interruption duration to cost per MW. In order to accurately calculate system expected LOLC, the LOLC coefficient needs to be evaluated according to the duration of each state (ω). We use mean duration of a state to calculate the cost coefficient. The cost coefficient can be improved further by assuming that the duration has exponential or some other distribution. The cost coefficient is then calculated as the expected value for different possibilities.

Mean duration of each stage can be calculated by the reciprocal of the equivalent transition rate from that state to others. State mean duration is presented in (12.14). The equivalent transition rate of all components can be calculated using the recursive formula in [14] when constructing the probability distribution function.

$$D_{\omega} = \frac{1}{\sum_{i \in I} \lambda_{gi}^{\omega+} + \sum_{i \in I} \lambda_{gi}^{\omega-} + \sum_{i,j \in I, i \neq j} \lambda_{lij}^{\omega+} + \sum_{i,j \in I, i \neq j} \lambda_{lij}^{\omega-} + \sum_{l \in L} \lambda_{l}^{\omega}} \quad (12.14)$$

D_{ω}: Mean duration of state ω in hours;

$\lambda_{gi}^{\omega+}$: Equivalent transition rate of generation in area i from a capacity of state ω to higher capacity in per hour;

$\lambda_{gi}^{\omega-}$: Equivalent transition rate of generation in area i from a capacity of state ω to lower capacity in per hour;

$\lambda_{lij}^{\omega+}$: Equivalent transition rate of transmission line from area i to area j from a capacity of state ω to higher capacity in per hour;

$\lambda_{lij}^{\omega-}$: Equivalent transition rate of transmission line from area i to area j from a capacity of state ω to lower capacity in per hour;

λ_{l}^{ω}: Equivalent transition rate of area load from state ω to other load states in per hour.

The customer damage function used in this chapter is taken from [85]; however, other damage functions, if known, could be used. The function was estimated from a electric utility cost survey in the US. For residential loads, interruption cost in dollars per kW-h can be described, as a function of outage duration, by (12.15):

$$c_{li}(\omega) = e^{0.2503 + 0.2211 D_{\omega} - 0.0098 D_{\omega}^2}. \quad (12.15)$$

Reliability-Constrained Consideration
We include a reliability consideration in the expansion problem by limiting the expected loss of load up to a pre-specified value, as shown in (12.16):

$$E_{\tilde{\omega}}\{y_{li}(\omega)\} \leq \alpha, \quad (12.16)$$

where α is the upper limit of expected load loss. This reliability constraint, together with a budget constraint, may cause the problem to be infeasible. Instead of directly imposing the reliability constraint, the objective function is modified using Lagrangian relaxation:

$$\min \sum_{i \in I} c_i x_i + E_{\tilde{\omega}}\{f(x, \tilde{\omega})\} + P \times (E_{\tilde{\omega}}\{y_{li}(\omega)\} - \alpha), \quad (12.17)$$

where P is a penalty factor if the expected load loss violates the limit α. Due to the budget constraint, it is possible that the resulting expected load loss is higher than the upper limit.

12.3 Sample Average Approximation (SAA)

The expected cost of load loss can be approximated by means of sampling. Let $\omega_1, \omega_2, \ldots, \omega_N$ be N realizations of random vector for all uncertainties in the model; the expected cost of load loss can be replaced by (12.18):

$$\tilde{f}_N(x) = \frac{1}{N} \sum_{k=1}^{N} f(x, \omega_k). \tag{12.18}$$

This function is a SAA of the expected cost of load loss. The problem can then be transformed into deterministic equivalent model as follows:

$$\min \sum_{i \in I} c_i x_i + \frac{1}{N} \sum_{k=1}^{N} \left\{ \sum_{i \in I} \left\{ c_{li}(\omega_k) y_{lik}(\omega_k) + c_{oi}(\omega_k) y_{gik}(\omega_k) \right\} \right\} \tag{12.19}$$

$$s.t. \sum_{i \in I} c_i x_i \leq B \tag{12.20}$$

$$\frac{1}{N} \sum_{k=1}^{N} y_{lik}(\omega_k) \leq \alpha. \tag{12.21}$$

For all $k \in \{1, 2, \ldots, N\}$,

$$y_{gik}(\omega_k) \leq g_i(\omega_k) + A_i x_i, \forall i \in I \tag{12.22}$$

$$b_{ij}(\theta_{ik}(\omega_k) - \theta_{jk}(\omega_k)) \leq t_{ij}(\omega_k), \forall i, j \in I, i \neq j \tag{12.23}$$

$$y_{lik}(\omega_k) \leq l_{ik}(\omega_k), \forall i \in I \tag{12.24}$$

$$\sum_{j \in I} B_{ij} \theta_{jk}(\omega_k) + y_{gik}(\omega_k) + y_{lik}(\omega_k) = l_{ik}(\omega_k), \forall i \in I \tag{12.25}$$

$$x_i \geq 0, \text{integer} \tag{12.26}$$

$$y_{gik}(\omega_k), y_{lik}(\omega_k) \geq 0, \forall i \in I \tag{12.27}$$

$$\theta_i(\omega_k) \quad \text{unrestrict.} \tag{12.28}$$

Note that, by virtue of the nature of sampling, a solution obtained from this sample-based approach does not necessarily guarantee optimality in the original problem. The optimal sample-based solutions, when obtained with different sample sets, instead provide statistical inference of a confidence interval of the actual optimal solution.

The reliability-constrained problem can be formulated by modifying the objective function with constraint (12.21) according to (12.17). Equation (12.13) can also be modified to incorporate unit availability as follows. For each scenario k,

$$y_{gik}(\omega_k) \leq g_i(\omega_k) + \sum_{q \in Q_i} A_{iq}(\omega_k) x_{iq}, \forall i \in I. \tag{12.29}$$

Let x_N^* be the optimal solution and z_N^* be the optimal objective value of an approximated problem. Generally, x_N^* and z_N^* vary by the sample size N. If x^* is the optimal solution and z^* is the optimal objective value of the original problem, then, obviously,

$$z^* \leq z_N^*. \tag{12.30}$$

Therefore, z_N^* constitutes an upper bound of the optimal objective value. Since z_N^* is the optimal solution of the approximated problem, then the following is true:

$$z_N^* = z_N^*(x_N^*) \leq z_N^*(x^*). \tag{12.31}$$

Taking expectation on both sides, (12.31) becomes:

$$E[z_N^*(x_N^*)] \leq E[z_N^*(x^*)]. \tag{12.32}$$

Since the SAA is an unbiased estimator of the population mean,

$$E[z_N^*(x_N^*)] \leq E[z_N^*(x^*)] = z^*, \tag{12.33}$$

which constitutes a lower bound of the optimal objective value. In the following, details on obtaining lower bound and upper bound estimates are discussed. The derivation of lower and upper bound confidence intervals was presented in [86] and has been applied in [18, 19].

Lower-Bound Estimates
The expected value of z_N^*, $E[z_N^*]$ can be estimated by generating M_L independent batches, each of N_L samples. For each sample set s, solve the SAA problem which gives $z_{N_L^s}$ and the lower bound can be found from (12.34):

$$L_{N_L,M_L} = \frac{1}{M_L} \sum_{i=1}^{M_L} z_{N_L,i}^*. \tag{12.34}$$

By the central limit theorem, the distribution of the lower-bound estimate converges to a normal distribution $\mathcal{N}(\mu_L, \sigma_L^2)$ where $\mu_L = E[z_{N_L}^*]$, which can be approximated by a sample mean L_{N_L,M_L}, and σ_L^2 can be approximated by a sample variance:

$$s_L^2 = \frac{1}{M_L - 1} \sum_{i=1}^{M_L} (z_{N_L,i}^* - L_{N_L,M_L})^2. \tag{12.35}$$

The two-sided $100(1 - \beta)\%$ confidence interval of the lower bound is found from (12.36):

$$\left[L_{N_L,M_L} - \frac{z_{\beta/2} s_L}{\sqrt{M_L}}, L_{N_L,M_L} + \frac{z_{\beta/2} s_L}{\sqrt{M_L}} \right], \tag{12.36}$$

where $z_{\beta/2}$ satisfies $\Pr\{z_{-\beta/2} \leq \mathcal{N}(0,1) \leq z_{\beta/2}\} = 1 - \beta$. It should be noted that the lower-bound confidence interval is computed by solving M_L independent SAA problems of sample size N_L.

Upper-Bound Estimates

Given a sample-based solution x_N^*, the upper bound of the actual optimal objective can be estimated by generating M_U independent batches, each of N_U samples. Since the solution is set to x_N^*, (12.19) can be decomposed based on system state ω to N_U independent linear programming (LP) problems. For each sample batch s, solving the LP problems gives $z_{N_U,i}^*(x_N^*)$. Then, the upper bound is approximated using equation (12.37):

$$U_{N_U,M_U}(x_N^*) = \frac{1}{M_U} \sum_{i=1}^{M_U} z_{N_U,i}^*(x_N^*). \tag{12.37}$$

By the central limit theorem, the distribution of the upper-bound estimate converges to a normal distribution $\mathcal{N}(\mu_U, \sigma_U^2)$ where $\mu_U = E[z_{N_U,i}^*(x_N^*)]$, which can be approximated by a sample mean U_{N_U,M_U}, and σ_U^2 can be approximated by a sample variance:

$$s_U^2(x_N^*) = \frac{1}{M_U - 1} \sum_{i=1}^{M_U} (z_{N_U,i}^*(x_N^*) - U_{N_U,M_U}(x_N^*))^2. \tag{12.38}$$

The two-sided $100(1 - \beta)\%$ confidence interval of the lower bound is found from (12.39):

$$\left[U_{N_U,M_U}(x_N^*) - \frac{z_{\beta/2} s_U}{\sqrt{M_U}}, U_{N_U,M_U}(x_N^*) + \frac{z_{\beta/2} s_U}{\sqrt{M_U}} \right], \tag{12.39}$$

where $z_{\beta/2}$ satisfies $\Pr\{z_{-\beta/2} \leq \mathcal{N}(0,1) \leq z_{\beta/2}\} = 1 - \beta$.
A solution x_N^* is found from each batch s of M_L batches in lower-bound SAA problems and used to compute the upper-bound estimates. It should be noted that the upper-bound confidence interval depends on the chosen approximate solution x_N^* from SAA problems. Thus, M_L upper-bound intervals are computed.

Optimal Solution Approximation

An optimal solution can be extracted when a unique solution is obtained from solving several SAA problems with different samples of a given size, N. In theory, optimality should be attained with sufficiently large N. This means that it may be possible that each sample yields different solutions for small sample size. If an identical solution is found from solving SAA problems with these samples, it is highly likely that the solution is optimal or close to optimal. In addition to obtaining identical solutions, confidence intervals of the lower bound and upper-bound estimates are also used to validate the approximate solution. If the intervals of both lower and upper-bound estimates are close enough, then the approximate solution tends to be close to the optimal solution. The SAA offers a solution to the large-scale problems that are otherwise computationally intractable. The optimal solution validation is still a fairly open research topic. The number of samples, as well as number of batches, also plays an important part. Large sample sizes increase the computation burden, while small sample sizes do not seem to represent the entire state space well. Interested readers are referred to [18, 19, 86] for more information about the optimal solution verification.

12.4 Computational Results

We report the results of three studies in this section. The first study is to determine optimal generation planning solution where the lower and upper bounds of the objective functions are estimated. The second study is to incorporate unit availability in the generation expansion problem. The third study is to analyze the performance of system-wide reliability maximization. The test system of all three studies is the 24-bus IEEE-RTS. The generator and transmission line parameters can be found in [35]. In order to reduce the number of system states, system load is grouped into 20 clusters using a clustering algorithm. Cost of additional units is assumed and shown in Table 12.1. Five buses are chosen as possible locations for adding units. With original load, the expected load loss (α) is 0.07 MW, and this is used as the specified limit in the optimization.

Table 12.1 Additional Generation Parameters

Bus	Unit capacity (MW)	Cost ($m)
101	20	20
102	20	20
107	100	100
115	12	12
122	50	50

The penalty factor (P) of 10^6 is used. The load is increased by 10% to represent projected demand growth. The budget is $100 million dollars. The total number of system states $(|\Omega|)$ of this problem is 9×10^{18}. To simplify the problem, operation cost in the second-stage objective function is neglected. This is also due to the lack of data for operation cost of existing units in IEEE-RTS. However, there is no inherent limitation in the methodology to include the operation cost in the problem.

12.4.1 Optimal Generation Planning Problem

To compare the effectiveness of the SAA using Monte Carlo sampling, four different sample sizes are chosen, namely, 500, 1000, 2000 and 5000. The lower-bound estimate of each sample size is calculated by solving SAA problems with data generated by five different batches of the sample. Therefore, M_L is 5 and N_L are 500, 1000, 2000 and 5000. Note that, at this point, each sample size will produce five solutions, which may or may not be identical, from five batches of sample. These solutions are then used to calculate the upper-bound estimate. The upper-bound estimate of each sample size is obtained by substituting the solution obtained from that particular SAA problem. This will transform a SAA problem into a independent linear programming problem, which makes it faster to solve a than SAA problem. In this study, five batches of sample of size 10000 are used to estimate the upper bound. Thus, M_U is 5 and N_U is 10000.

The 95% confidence intervals of the lower bound from different sample sizes, and the 95% confidence intervals of upper bound from different batches of sample size (N_L) are shown in Figure 12.1. For each sample size, the best upper-bound estimate is chosen from the tightest confidence interval. If the interval is the same, the best upper bound is found from minimum average value.

The lower-bound intervals overall tend to be higher and tighter when sample size increases, except for sample size of 5000. This is due to the nature of sampling. When sample sizes are smaller, such as 500 and 1000, the lower-bound intervals may be higher than the upper bound. This may be due to the fact that the upper bound is found from the sample size of 10000. When a sample size is small, the duration of sampled states may be overestimated, which results in higher cost coefficient and expected loss of load cost.

The solution obtained from different batches of sample size can be found from Table 12.2. An optimal solution is approximated by solving SAA problems with increased sample sizes. In this study, the optimal solution is assumed to be reached when identical solutions are found within five consecutive batches of sample of the same size. It can be seen from Table 12.2 that the solutions are identical when sample sizes are 5000. Therefore, the solution to this problem is to install 5 units at bus 102.

Even though the original problem has a large number of system states (9×10^{18}), SAA requires only a small, manageable sample size of 5000 to solve the optimization problem. With the budget of $100 million dollars, the optimal

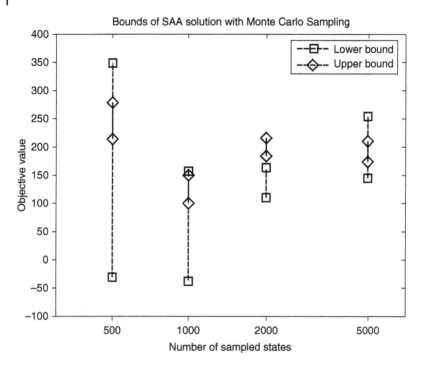

Figure 12.1 Bounds of SAA solution.

solution yields expected load loss of 0.08 MW. Note that this expected load loss is greater than the initial limit of 0.07 MW. This is due to the specified budget constraint.

12.4.2 Availability Considerations of Additional Units

In this analysis, a two-state Markov model is assumed to represent unit availability of additional generators using forced outage rate data of the IEEE-RTS. The formulation is rather general in that any three-or-more-state Markov model can be accommodated. With the same budget of $100 million dollars, the maximum number of additional generating units in each bus is shown in Table 12.3. The total number of system states ($|\Omega|$) of this problem is 1.9×10^{25}, compared to 9×10^{18} without availability consideration of additional units. The optimal solution approximation is found from four different sample sizes, which are 1000, 2000, 8000 and 12000, using Monte Carlo sampling.

The lower-bound estimate of the objective function from each sample size is calculated by solving SAA problems with data generated by four different batches of sample. Therefore, M_L is 4 and N_L are 1000, 2000, 8000 and 12000. Note that, at this point, each sample size will produce four solutions, which may or may not be identical, from four batches of sample. The 95% confidence

Table 12.2 Approximate Solutions

Sample Size	Batch	Number of Additional Units at Bus				
		101	102	107	115	122
500	1	0	0	0	0	0
	2	0	0	0	0	0
	3	0	5	0	0	0
	4	0	0	1	0	0
	5	0	1	0	0	0
1000	1	0	5	0	0	0
	2	0	2	0	0	0
	3	0	0	0	0	0
	4	0	0	0	0	0
	5	0	0	0	0	0
2000	1	0	5	0	0	0
	2	4	0	0	1	0
	3	0	5	0	0	0
	4	0	5	0	0	0
	5	0	5	0	0	0
5000	1	0	5	0	0	0
	2	0	5	0	0	0
	3	0	5	0	0	0
	4	0	5	0	0	0
	5	0	5	0	0	0

Table 12.3 Unit Availability Data of Additional Generation Parameters

Bus	Unit Capacity (MW)	Forced Outage Rate	Cost ($m)	Number of Units
101	20	0.1	20	5
102	20	0.1	20	5
107	100	0.04	100	1
115	12	0.02	12	8
122	50	0.01	50	2

Table 12.4 Lower-Bound Estimates

Sample Size	Objective Function Value	
	Expansion Cost ($m)	Expected Power Loss (MW)
1000	70 ± 37.53	0.0249 ± 0.0455
2000	100	0.1083 ± 0.0301
8000	100	0.0812 ± 0.0373
12000	100	0.0720 ± 0.0169

intervals of the lower bound of the objective function from different sample sizes are shown in Table 12.4.

The objective function is expansion cost and expected power loss, which are shown separately. The solution obtained from different batches of sample size can be also seen from Table 12.5.

Table 12.5 Approximate Solutions with Availability Considerations of Additional Units

Sample Size	Batch	Number of Additional Units at Bus				
		101	102	107	115	122
1000	1	1	4	0	0	0
	2	0	3	0	0	0
	3	0	5	0	0	0
	4	0	1	0	0	0
2000	1	0	5	0	0	0
	2	1	4	0	0	0
	3	2	3	0	0	0
	4	3	2	0	0	0
8000	1	1	4	0	0	0
	2	2	3	0	1	0
	3	2	3	0	0	0
	4	1	4	0	0	0
12000	1	2	3	0	0	0
	2	2	3	0	0	0
	3	2	3	0	0	0
	4	2	3	0	0	0

It can be seen from Table 12.5 that the solutions are identical when sample sizes are 12000. Therefore, the solution of the problem with availability considerations of additional units is to install 2 units at bus 101 and 3 units at bus 102, which yields expected power loss of 0.099 MW. This study shows that unit availability consideration can affect the solution of the generation expansion problem.

12.4.3 Comparative Study of System-Wide Reliability-Constrained Generation Expansion Problem

Generation and peak load data at each bus are shown in Table 12.6. All reliability indices are found from Monte Carlo simulation with a coefficient of variation of 0.05. Loss-sharing policy is implemented in this study. We ignore unit availability consideration in this case study.

Table 12.6 Generation and Peak Load

Bus	Generation (MW)	Peak load (MW)
101	192	118.8
102	192	106.7
103	-	198
104	-	81.4
105	-	78.1
106	-	149.6
107	300	137.5
108	-	188.1
109	-	192.5
110	-	214.5
113	591	291.5
114	-	213.4
115	215	348.7
116	155	110
118	400	366.3
119	-	199.1
120	-	140.8
121	400	-
122	300	-
123	660	-

Table 12.7 Expected Unserved Energy before and after Optimal Planning in MWh Per Year

Bus	Before Optimal Planning	After Optimal Planning
103	41.50	17.91
104	47.23	27.76
107	13.59	5.86
108	2.50	0.31
109	1022.06	591.24
110	9.23	5.08
113	0.00	0.88
114	6.39	6.25
115	33.27	21.51
116	3.95	2.27
118	20.48	17.55
119	0.00	0.61
120	0.00	1.04

An optimal solution, found in the first case study, is to install 5 units at bus 102, which yields system expected power loss of 0.08. Expected unserved energy of each area before and after optimal planning is shown in Table 12.7.

It can be seen from Table 12.7 that the system-wide reliability maximization planning does not necessarily yield equally distributed reliability level in each bus. However, it does tend to improve the reliability at almost all buses, especially those with a high level of unserved energy. In some cases, buses 113, 119 and 120 sacrifice their reliability for the system since the reliability level of these buses are worse after the optimal planning. The EUE of these buses is, however, very small to start with. EUE percentage reduction is shown in Table 12.8.

The percentage EUE reductions of each bus vary from 2% to 87%. Most of the buses with large EUE reductions, for example, buses 103–110, are in the neighboring area with the additional units at bus 102. Other distant buses from bus 102 seem to have smaller reliability improvement. The results indicate that though the reliability improvement may not be proportionately distributed across buses, but most of the buses do experience improvement, and the most unreliable ones experience a considerable gain.

12.5 Conclusion and Discussion

A stochastic programming approach with sample average approximation is presented for the composite-system generation adequacy planning problem.

Table 12.8 Expected Unserved Energy Reduction

Bus	EUE Reduction in MWh Per Year	Percentage Reduction (%)
103	23.58	56.82
104	19.47	41.22
107	7.73	56.88
108	2.19	87.60
109	430.81	42.15
110	4.15	44.96
113	−0.88	-
114	0.14	2.19
115	11.75	35.32
116	1.69	42.78
118	2.93	14.31
119	−0.61	-
120	−1.04	-

The problem is formulated as a two-stage recourse model with the objective to minimize expansion cost in the first-stage and operation and reliability cost in the second-stage. Reliability is included in terms of expected cost of load loss in the objective function and expected load loss in the constraint. Availability of additional units can also be incorporated in the formulation, which may result in a exponentially large number of system states.

Due to numerous system states, straightforward implementation of the L-shaped method is impractical, if not impossible. To overcome this, exterior sampling method is proposed. The reliability function of the problem is approximated by the sample-average using Monte Carlo sampling. Generation expansion planning is implemented on the 24-bus IEEE-RTS. Results show that even though the problem itself has a huge number of system states, the proposed method can effectively estimate the optimal solution with a relatively small number of samples. The planning problem includes system reliability considerations, which may be of interest to Independent System Operators when designing price incentive programs for the generation companies.

A comparative study on the customer reliability level at each bus before and after optimal planning is conducted. Results show that system-wide reliability optimization may not equally improve the reliability level at each bus but most buses experience improvement in reliability, especially those suffering the most. It is likely that buses within a close distance to the additional generators benefit more from the optimal planning. Perhaps that is the reason for the placement of the additional generators in those locations. This information is useful for Independent System Operators in discharging their responsibility to oversee the electric market and design appropriate mechanisms in order to promote fair pricing for reliability.

13

Modeling of Variable Energy Resources

13.1 Introduction

Since the beginning of this century we have witnessed an acceleration in the adoption of renewable energy resources and technologies. Various forces— social, political, economic, regulatory and technological—have contributed to the creation of a climate that fosters the development and proliferation of numerous technologies that enable the conversion, control and integration of renewable energy resources. Although the mix of renewable resources is diverse, ranging from wind and solar to tidal and biomass, the bulk of recent investments have gone into wind and solar, both of which are considered variable resources because their availability is usually subject to the whims of nature.

It is important to suitably model these variable resources for grid reliability analysis and planning. It should be recognized that these resources behave differently from conventional resources that are assumed to have continuous and unlimited supply of fuel or other underlying resources, and therefore possess the following important characteristics from the perspective of reliability modeling and evaluation.

1. *Dispatchability*, i.e., the ability to be committed and dispatched as part of operational planning procedures.
2. *Independence*, i.e., they can be assumed to have generating models that are statistically independent of each other.

This chapter discusses how the absence of these characteristics affects the performance of variable energy resources (VERs) and how VERs have been modeled so as to include them in grid reliability analysis and planning. It is important to note that this field is undergoing considerable development at the present time and that models described here are likely to evolve in the near future.

Electric Power Grid Reliability Evaluation: Models and Methods, First Edition. Chanan Singh, Panida Jirutitijaroen, and Joydeep Mitra.

13.2 Characteristics of Variable Energy Resources

As discussed, VERs are not dispatchable. Therefore, in developing reliability models for VERs it is important to include two stochastic elements: forced outages of the generating components, such as wind turbines and solar panels, and the stochasticity in the underlying natural resources, such as wind and solar radiation. Most established models assume these two elements to be statistically independent. Although recent discoveries [87] have challenged these assumptions in some cases, the majority of present-day models are based on this assumption of independence. As more experience is acquired, these dependencies too will be accounted for; as already stated, this area is evolving.

Another aspect of VER plants is that they appear in *cohorts*, e.g., wind turbine generators (WTG) within a wind farm and panels in a solar park, wherein the constituent components show a strong correlation in output as a result of being subject to the same underlying resource—wind speed or insolation. Consequently, they cannot be modeled using the approach described in Chapter 8. The practice has been to model the entire cohort (wind farm or solar park) as a single equivalent unit with multiple states. Modeling of correlations is complex and merits further discussion.

13.2.1 Correlation within a VER Cohort

Most correlation studies have focused on wind generation because of the larger variability in wind speeds and complex interactions within wind farms. Several studies have investigated the effect of the correlation between the output power of WTGs and the change in the wind speed within the same farm [88–92]. The degree of the correlation depends on several factors, such as geographic location, separation distances between WTGs, terrain and number of WTGs on the wind farm, and their influence on wake effects and turbulence [93–95]. These correlations also vary with time scales (the frequency of wind speed observations can be intervals of 5 minutes, 10 minutes, hourly, etc.). Thus, determining the exact degree of correlation between individual WTGs and the change in the wind speed within the same farm is a challenging and complex task, particularly for large wind farms [96]. However, it has been reported that WTGs within a wind farm are approximately driven by same values of wind speed, and the power outputs of WTGs show consistent behavior among individual WTGs with the change in the wind speed on the entire farm [89–92, 97]. Hence, similar WTGs will produce similar outputs with some dispersion; these outputs can be represented by a cluster, with the cluster mean representing their average output. This notion is used in the models presented in this chapter.

13.2.2   Correlation with Other Stochastic Elements

Correlations have also been observed between VER production and other stochastic elements that are considered in power system analyses. For instance, [98] reports negative correlations between wind and solar production, and [99] reports positive, negative and negligible correlations between wind production and system loads depending on time scales, time of day and seasons. While some researchers [100, 101] have modeled the correlation with load, others [102] have postulated that the correlation is negligible. On the other hand, correlations between solar production and loads have been found to be positive [103]. Moreover, the impact of correlation is also affected by such factors as differences in scale (sizes of wind or solar installations and load), geographic location and electrical distances between points of insertion.

Another challenge in modeling correlation between VER production and system load is that most of these models treat VER injections as negative loads, inherently assuming that all energy that can be produced by a VER cohort is completely accepted by the system. Despite VER production being designated as "must take" in many regulatory regimes, it is well known that prevailing system conditions, such as low loads or network congestion, often necessitates spillage, particularly that of wind generation.

In view of the above reasons, most models in the literature fall into two categories—those that include correlations with load [100] and those that do not [104].

13.3   Variable Resource Modeling Approaches

We describe here some typical approaches that have been used for modeling cohorts of VERs. In these approaches, the transmission system is assumed capable of transporting generation to load points, i.e., the transmission constraints are ignored. The analysis at the composite-system level is discussed in Section 13.4.

13.3.1   Approach I: Capacity Modification Method

In this approach [105, 106], the entire power system is divided into several subsystems. While one of these corresponds to all the conventional units combined, the others correspond to the different types of unconventional units. For simplicity's sake, the approach is explained using a sample system consisting of a conventional and two unconventional subsystems. The unconventional subsystems consist of units being operated on solar and wind energy, respectively. General expressions for calculating the LOLE and EUE indices for a system consisting of n number of unconventional subsystems can, however, be found in [107].

For each subsystem, a generation system model is developed using the unit addition algorithm [108]. To start with, the unconventional units are treated in a conventional manner in the sense that the traditional two-and three-state unit models [9] are used. Also, full capacities for each state of the unconventional units are used.

The next step is to create vectors containing the hourly power outputs of the unconventional subsystems such that the term $POU_{l,k}$ represents the power output of the lth unconventional subsystem during the kth hour of the period of study [105]. The vector corresponding to a given unconventional subsystem is then divided by its rated power output in order to obtain the weight vector, which indicates the fraction of the total rated unconventional power that is being effectively generated at the various hours of study. This can be mathematically expressed as:

$$A_l = (l/PRU_l)[POU_{l,1}POU_{l,2}POU_{l,3}...POU_{l,N_t}]. \tag{13.1}$$

In (13.1), the terms A_l and PRU_l respectively represent the weight vector and the rated power output corresponding to the lth unconventional subsystem, and $POU_{l,k}$ is the actual power output of the lth unconventional subsystem for the kth hour. N_t is the number of hours in the study period.

To incorporate the effect of fluctuating energy, the generation system models of the unconventional subsystems are modified hourly depending on their energy output levels. This is achieved by multiplying the rated generation capacity vector **CU** of a given unconventional subsystem (say l) with the term $A_{l,k}$, where $A_{l,k}$ represents the fraction of the total rated unconventional power produced by the lth subsystem for the kth hour of study. It should, however, be noted that these hourly modifications do not affect the state probability vectors of the respective generation system models. The models corresponding to all the subsystems are then combined hourly using a discrete state method in order to calculate the LOLE and EUE indices.

Discrete State Method

We shall explain this method [105] using our sample system consisting of a conventional and two unconventional subsystems. Let us now define the following vectors associated with the three generation system models as:

CC, $\widehat{PC}$ = Generation capacity and cumulative probability vectors associated with the model corresponding to the conventional subsystem.

CU_1, $\widehat{PU_1}$ = Generation capacity and cumulative probability vectors associated with the model corresponding to the lth unconventional subsystem, where $l \in [1, 2]$.

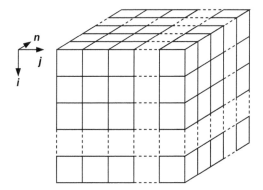

Figure 13.1 State space diagram for the combination of three subsystems.

Since each of these subsystems is treated as a multistate unit, the combination of their generation system models for the kth hour of study results in distinct states with capacities given by [105]:

$$C_{ijn,k} = CC_i + A_{1,k}CU_{1,j} + A_{2,k}CU_{2,n}. \tag{13.2}$$

In (13.2), the subscripts i, j and n refer to the different states in the first, second and third subsystem, respectively. The state space diagram for this combination can be represented by a cuboid, as shown in Figure 13.1.

The next step is to find the boundary states that define a loss of load and compute the corresponding probability and expected unserved energy. Looking at Figure 13.1, one may observe that for a given value of n there exists a two-dimensional state space for different values of i and j. The boundary of capacity deficiency in this state space can then be found by varying i for each value of j, until $C_{ijn,k}$ is found to be less than or equal to the load for the hour in question [105]. The loss of load probability (LOLP) computed along a given $i - j$ state space for the kth hour of study can thus be expressed as:

$$\text{LOLP}_{k,n} = \sum_{j=1}^{nu1} \widehat{PC}_{th}(\widehat{PU}_{1,j} - \widehat{PU}_{1,j+1})(\widehat{PU}_{2,n} - \widehat{PU}_{2,n+1}). \tag{13.3}$$

In (13.3), the subscript th denotes a threshold and is valid only for given values of j and n. It is numerically equal to the smallest value of i such that the expression $(CC_{th} + A_{1,k}CU_{1,j} + A_{2,k}CU_{2,n}) \leq L_k$ is satisfied for the k_{th} hour of study. L_k is the system load level for the k_{th} hour, and $nu1$ represents the total number of states in the first unconventional subsystem. The LOLP for the given hour in question can then be calculated by summing up (13.3) over all values of n. Thus:

$$\text{LOLP}_k = \sum_{n=1}^{nu2} \text{LOLP}_{k,n}. \tag{13.4}$$

In (13.4), the term $nu2$ refers to the total number of states in the second unconventional subsystem. The loss of load expectation (LOLE) for the kth hour of study, LOLE_k, can be computed by multiplying LOLP_k by the duration of time step. LOLE_k is numerically equal to LOLP_k in this case, as the time step is one hour. Finally, the LOLE for the entire period of study is obtained as:

$$\text{LOLE} = \sum_{k=1}^{N_t} \text{LOLE}_k \tag{13.5}$$

The expected unserved energy (EUE) can be computed using (8.45) and (8.46) by constructing a system negative margin table for each hour of the period of study, and by noting that ΔT for our case is equal to 1 hour. A detailed algorithm for implementing this proposed approach and relevant equations for calculating the hourly power outputs of solar power plants and wind turbine generators are given in [105].

13.3.2   Approach II: Clustering Method

The capacity modification approach described in Section 13.3.1 yields accurate values of the system reliability indices but is computationally inefficient owing to the hourly calculations involved. It is particularly unsuitable for calculating EUE, as the hourly construction of a system negative margin table, and the subsequent computations drastically increases the CPU time. The clustering approach was therefore proposed in [109] for efficiently calculating the reliability indices with minimum computational effort.

Clustering, or grouping, is done on the basis of similarities or distances between data points. The inputs required are similarity measures or data from which similarities can be computed. The study conducted in [109] was based on the FASTCLUS [110] method of clustering owing to its suitability for use on large data sets. The central idea of the FASTCLUS approach is to choose some initial partitioning of the data units and then alter cluster memberships so as to obtain a better partition. FASTCLUS is designed for performing a disjoint cluster analysis on the basis of Euclidian distances computed from one or more variables. The observations are partitioned into clusters such that every observation belongs to one and only one cluster. A detailed description of the FASTCLUS clustering method can be found in [37]. The relevant algorithm for implementing this method and an example demonstrating how to use it for grouping some random observations into two clusters are given in [109].

The first few steps of this approach are similar to those of Approach I in the sense that the entire system is again divided into several subsystems corresponding to the conventional and the different types of unconventional units. A generation system model is then built for each such subsystem. The next step is to compute the hourly power outputs of the unconventional subsystems and define a set of N_t different vectors as described below:

$$d_k = [(L_k/L_{peak})A_{1,k}A_{2,k}, ...A_{l,k}], k \in [1, 2, ...N_t].$$ (13.6)

All terms used in (13.6) are as described in Section 13.3.1. L_{peak} refers to the peak load for the entire period of study. Equation (13.6) basically aims to model the correlation between the hourly load and the fluctuating power outputs of the unconventional subsystems. Using the FASTCLUS method, the set of vectors defined by equation (13.6) are then grouped into different clusters. Assuming that the cth cluster contains a set of v vectors (as defined by (13.6)), it's "centroid" is defined as the mean of all these vectors and is in turn denoted by the vector d^c. Thus:

$$d^c = [(L^c A_1^c A_2^c ...A_l^c], c \in [1, 2, ...N_c].$$ (13.7)

In (13.7), while the term L^c refers to the mean value of the load in the cth cluster expressed as a fraction of the peak load, the term A_l^c refers to the mean value of the fraction of the total rated unconventional power that is effectively produced by the lth unconventional subsystem in the cth cluster. N_c refers to the total number of clusters chosen for a given simulation. At the end of the simulation, the FASTCLUS routine thus outputs the following parameters for each cluster: *Frequency* (the total number of vectors d_k belonging to a given cluster) and the *Cluster Centroid* (the vector d^c for that cluster) [109].

To incorporate the effect of fluctuating energy, the generation system models of the unconventional subsystems are modified for each cluster. This is achieved by multiplying the rated generation capacity vectors of the various unconventional subsystems with the corresponding A_l^c terms derived from the respective clusters. The generation system models corresponding to all the subsystems are then combined for each cluster in order to calculate the relevant reliability indices. To illustrate this concept using our sample system (refer to Section 13.3.1), let us now rewrite (13.2) with respect to the cth cluster as follows:

$$C_{ijn}^c = CC_i + A_1^c.CU_{1,j} + A_2^c.CU_{2,n}.$$ (13.8)

Referring to our sample system, the loss of load probability for the cth cluster (LOLP^c) can be calculated using (13.3) and (13.4). One should note that the subscript k denoting a given hour in those two equations will now be replaced by the superscript c denoting a given cluster. Using the concept of conditional probability, the LOLP for the entire system can then be obtained as:

$$\text{LOLP} = \sum_{c=1}^{N_c} \text{LOLP}^c P(d^c). \tag{13.9}$$

In (13.9), the term $P(d^c)$ refers to the probability of occurrence of the cth cluster and is obtained by dividing the cluster's frequency by N_t. Finally, the LOLE for the entire period of study is obtained as:

$$\text{LOLE} = \text{LOLP} \times N_t. \tag{13.10}$$

The expected unserved load for the cth cluster, U^c, can be calculated using (8.46) by constructing a system negative margin table for the given cluster and by noting that the subscript k in the equation will now be replaced by the superscript c. The expected unserved energy for the cth cluster, EUE^c, can then be obtained by multiplying U^c with N_t. Using the concept of conditional probability, the EUE for the entire system is finally obtained as:

$$\text{EUE} = \sum_{c=1}^{N_c} \text{EUE}^c P(d^c). \tag{13.11}$$

A close look at Approaches I and II reveals that in the former, the modifications of the generation capacity vectors of the unconventional subsystems and the combination of the generation system models were carried out every hour; these operations are performed on a cluster-by-cluster basis in the latter. Since the number of clusters is typically much smaller than the number of hours under study, the clustering method is much more efficient. It should, however, be noted that the indices calculated using this approach are not as accurate as those obtained in [105], as the contents of the d_c vectors based on which the computations are performed for each cluster are obtained by averaging the corresponding values over a number of hours. This gives rise to some approximations in the calculations. The accuracy of the indices calculated using this method is a function of the number of clusters chosen for a given simulation. Certain techniques [111, 112] can, however, be used for choosing the optimum number of clusters. One should also note that if the number of clusters is equal to the number of hours in the study period, i.e., if N_c is equal to N_t, approaches I and II become identical to each other.

13.3.3 Approach III: Introduction of Mean Capacity Outage Tables

As described earlier, the reliability indices calculated using the clustering method introduce some approximation owing to the inherent approximations involved with clustering. Additionally, since the calculations are performed for each cluster, it is impossible to obtain the hourly contributions to the reliability indices using Approach II. The concept of *Mean Capacity Outage Tables* was therefore proposed in [107] for efficiently and accurately calculating EUE on an hourly basis. It should be pointed out at this time that this approach is essentially used for simplifying the EUE calculations only. The equations concerning the computation of LOLE are still the same as formulated in Section 13.3.1 (refer to (13.3) – (13.5)).

The first few steps of this approach are again similar to those of Approach I, in the sense that the entire system is divided into several subsystems corresponding to the conventional and the different types of unconventional units. A generation system model is then built for each such subsystem. To incorporate the effect of fluctuating energy, the generation system models of the unconventional subsystems are modified hourly depending on their energy output levels. The models corresponding to all the subsystems are then combined hourly in order to calculate the LOLE and EUE indices. Referring to our sample system described in Section 13.3.1, let us now define the following vectors in addition to those ($\mathbf{CC}, \widehat{\mathbf{PC}}, \mathbf{CU_l}, \widehat{\mathbf{PU_l}}$) already presented in Section 13.3.1:

$\mathbf{XC}$ = Capacity outage vector associated with the generation system model corresponding to the conventional subsystem.

$\mathbf{XU_l}$ = Capacity outage vector associated with the generation system model corresponding to the lth unconventional subsystem, where $l \in [1, 2]$.

We shall now rewrite (13.2) in terms of the system capacity outages as follows:

$$X_{ijn,k} = XC_i + A_{1,k} \times XU_{l,j} + A_{2,k} \times XU_{2,n}. \tag{13.12}$$

Let us also define the term "critical capacity outage" for the kth hour of study as (13.2):

$$X_k = CC_1 + \sum_{l=1}^{2}(A_{l,k} \times CU_{l,1}) - L_k. \tag{13.13}$$

All terms used in (13.13) are as described in Section 13.3.1. It may be noted that the expression $(CC_1 + \sum_{l=1}^{2}(A_{l,k} \times CU_{l,1}))$ represents the effective total generation capacity of the system during the kth hour of study. For a given hour, say k, a loss of load situation occurs when $X_{ijn,k} > X_k$ for given values of i, j and n. The expected unserved load during the kth hour of study, U_k, can now be expressed as follows [107]:

$$U_k = \sum_{X_{ijn,k} > X_k} ((X_{ijn,k} \times P(X_{ijn,k}))). \tag{13.14}$$

In (13.14), the term $P(X_{ijn,k})$ is used to represent the probability that a system capacity outage occurs exactly equal to $X_{ijn,k}$ MW. We shall now demonstrate how the hourly computation of the system negative margin table for calculating U_k can be avoided by the application of a mean capacity outage table. The LOLP for the kth hour of study, LOLP_k can be expressed as [107]:

$$\text{LOLP}_k = \sum_{X_{ijn,k} > X_k} P(X_{ijn,k}). \tag{13.15}$$

Let H_k represent the expected (mean) value of all system capacity outages that would cause capacity deficiency during hour k [107]. Thus:

$$H_k = \sum_{X_{ijn,k} > X_k} (X_{ijn,k} \times P(X_{ijn,k})). \tag{13.16}$$

Using (13.15) and (13.16), we can rewrite (13.14) as [107]:

$$U_k = H_k - X_k \times \text{LOLP}_k. \tag{13.17}$$

While the term X_k in (13.17) can be calculated using (13.13) for a given hour k, LOLP_k can be computed using (13.3) and (13.4). Let us now expand (13.16) by using relevant terms from (13.3), (13.4) and (13.12):

$$H_k = \sum_{n=1}^{nu2} \sum_{j=1}^{nu1} \sum_{i=th}^{nc} [(XC_i + A_{1,k} \times XU_{1,j} + A_{2,k} \times XU_{2,n}) \times PC_i \times PU_{1,j} \times PU_{2,n}]. \tag{13.18}$$

It should be noted that the terms PC and PU in (13.18) refer to the respective state probabilities and not the cumulative probabilities as was the case in (13.3). nc represents the total number of states in the conventional subsystem. The term th in (13.18) can now be redefined in terms of the system capacity outage as the smallest value of i for which the expression $(XC_i + A_{1,k} \times XU_{1,j} + A_{2,k} \times XU_{2,n}) > X_k$ is satisfied for given values of j and n. Using the relevant notation for cumulative probability (refer to (13.3)), (13.18) can be rearranged as:

$$H_k = \sum_{n=1}^{nu2} \sum_{j=1}^{nu1} \{PU_{1,j} \times PU_{2,n} \times [A_{1,k} \times XU_{1,j} \times \widehat{PC_{th}} + A_{2,k} \times XU_{2,n} \times \widehat{PC_{th}}$$
$$+ \sum_{i=th}^{nc} (XC_i \times PC_i)]\}. \tag{13.19}$$

In order to simplify (13.19), we define [107]:

$$\widehat{HC(q)} = \sum_{i=q}^{nc} (XC_i \times PC_i). \tag{13.20}$$

Substitution of (13.20), with $q = th$, in (13.19) yields:

$$H_k = \sum_{n=1}^{nu2} \sum_{j=1}^{nu1} \{PU_{1,j} \times PU_{2,n} \times [A_{1,k} \times XU_{1,j} \times \widehat{PC_{th}} + A_{2,k} \times XU_{2,n} \times \widehat{PC_{th}}$$

$$+\widehat{HC_{th}}]\}. \tag{13.21}$$

We refer to the term $\widehat{HC(q)}$ for q = 1, 2, 3, nc, as the mean capacity outage table of the conventional subsystem. This table is the key concept proposed in [107] for efficiently computing EUE. Once the cumulative probability vector $\widehat{PC}$ associated with the generation system model of the conventional subsystem is computed, the construction of the mean capacity outage table $\widehat{HC(q)}$ requires little additional computational effort, as one can use a simple recurrence relation [113]. The expected unserved load during hour $k(U_k)$, as expressed in (13.17), can therefore be calculated using (13.3), (13.4), (13.13) and (13.21). The EUE for the entire period of study is then finally calculated by summing U_k over all hours.

The advantage of using (13.17) over (8.46) for calculating U_k can be realized by observing that the use of the mean capacity outage table essentially eliminates the need for carrying out hourly computations of a system negative margin table, thereby saving considerable simulation time. The relevant algorithm for implementing this proposed approach, as well as an example demonstrating how to use it on a sample system for calculating EUE, can be found in [106, 107].

13.4 Integrating Renewables at the Composite System Level

In Section 13.3, the variable energy resources were grouped into several categories, with the conventional generation all in one group, and the indices were computed for not being able to satisfy the total load. Thus it was assumed that the transmission was capable of transporting available generation to the load points. This is the typical generation planning model. As described in Chapter 11, at the composite-system level the constraints imposed by the capacities and failures of the transmission lines are considered. In this model, the generation (conventional and unconventional) and load are distributed over many buses that are connected together by the transmission lines. Some type of power flow calculation is needed to determine whether the loads can be satisfied given the state of generation and transmission. Although there are analytical methods to deal with the composite system reliability, it is generally agreed that Monte Carlo is the preferred method to deal with the complexity and size of the system.

Monte Carlo simulation broadly falls into sequential or nonsequential approaches, which is also called sampling. In Sections 13.4.1 and 13.4.2 we describe these two approaches for including the renewable resources.

13.4.1   Sequential Monte Carlo for Including Renewable Resources

The procedure for sequential simulation and its convergence for composite power systems is described in Section 11.4. The same procedure can be used while integrating the renewable energy resources with some modification in the core algorithm. The core algorithm is reproduced here with some modifications and described below:

1. Input system data (component state capacities and transition rates, dependencies, interconnection information, maintenance schedules and hourly load curve and hourly wind and solar data). Define convergence criteria for indices.
2. Initialize sample path index at $i = 1$.
3. Initialize component states at hour $j = 1$.
4. For prevailing component states at hour j, perform the following: for every component, draw a random number and determine the time to the next transition using one of the methods outlined in Section 6.3.1. Of these times, select the smallest, $t_{min}(k_m)$; the corresponding time and transition indicate the most imminent event, i.e., after t_{min}, the component k will transit to state m.
5. At hour j, perform the following:
   (a) Decrease by 1 hour the time t_{min} to the next event as determined in step 4.
      (i) If $t_{min} > 0$, no change has occurred in system states; proceed to step 5(b).
      (ii) If $t_{min} = 0$, update state of component k to m for hour j. If there are dependencies triggered by this transition, update states accordingly. For prevailing component states at hour j, perform the following: for every component, draw a random number and determine the time to the next transition. Of these times, select the smallest, $t_{min}(k_m)$. In other words, determine a new $t_{min}(k_m)$ as soon as a transition occurs.
      (iii) At hour j, is any component taken out for maintenance or brought back into service after maintenance? If so, alter its state accordingly.
   (b) For the prevailing component states as determined in step 5(a), do the following:
      (i) Update load at hour j
      (ii) Modify the capacities of renewable generation by the available wind velocity or solar insolation as the case may be.

(iii) Determine if the system experiences loss of load. If so, update estimates of indices as described in Section 6.4.2; otherwise proceed to step 6.

6. Is j equal to last hour in study period? If so, proceed to step 7; otherwise increase j by one hour and go to step 5.

7. Have the estimates of the indices converged to criteria specified in step 1? If so, end simulation and report results (index estimates and, optionally, convergence-related information, computation time, etc.); otherwise increment sample path index i by 1 and go to step 3.

The rest of the procedure is the same as in Chapter 11. In step 5(b)iii, a testing of system performance is involved. Different methods for testing of system states are discussed in Section 11.6; however, DC optimal flow is generally used.

13.4.2 Nonsequential Monte Carlo for Including Renewable Resources

The non-sequential Monte Carlo described in Section 11.5.1 can be adopted to include renewable energy sources as follows:

1. Input system data (component state capacities and probabilities, interconnection information, maintenance schedules and hourly bus loads, hourly wind speed and solar insolation at different bus locations). Define convergence criteria for indices.
2. Based on the maintenance schedules, divide the study period into intervals such that one or more components are taken out for maintenance or brought back from maintenance only at the beginning or end of an interval. In other words, the only outages that occur within an interval are forced outages and not planned outages. Based on the relative durations of the maintenance intervals, construct a discrete probability distribution function over all the maintenance intervals.
3. For every maintenance interval, identify the hourly loads and wind speed.
4. For every system component, construct a discrete probability distribution over the capacity states that the component can assume.
5. Initialize sample index $i = 1$ and sample size $N = 1$.
6. For the sample index i, draw a system state x_i as follows:
   (a) Generate a random number and use it to draw a maintenance interval.
   (b) Generate a random number and use it to draw a load level at each bus within the maintenance interval drawn in step 6(a). All hours have an equal chance in the sampling process. The bus load and bus wind speed related to the hour are used in the analysis. The wind speed for the hour is used to determine the power output of a wind or solar farm.

(c) For every component that is not on maintenance in the interval drawn in step 6(a), draw a random number and use it to determine the capacity of the component.

The combination of component capacities, load levels and wind speed or solar radiation sampled in steps 6(b) and 6(c) describes the system state x_i. Note that this manner of sampling ensures that the probability of drawing x_i equals the actual probability of the physical system assuming the corresponding state.

7. Determine if the system experiences loss of load in state x_i. If so, update estimates of indices and test for convergence as described in Section 6.4.1. If the indices have converged to the specified tolerance, then end the simulation and report results. Otherwise increment the sample index i and sample size N and go to step 6.

This algorithm serves to convey an idea of the mechanics of the nonsequential simulation method. Instead of using hourly load data, clustering can also be used in sampling [114]. First, the system load and power output of variable outputs are clustered using the *k-means* algorithm. Then instead of sampling an hour from the interval, one of these cluster agents is sampled to show the load and renewable states. In this method, the cluster agents are selected based on their probabilities. The probability of each cluster is a number in the range of [0, 1]. Thus, a roulette wheel is used to select cluster agents. This method may accelerate convergence of Monte Carlo simulations because system load and power output of renewables are sampled from a smaller set instead of all data units.

14

Concluding Reflections

In wrapping up the subject of reliability and the models and methods applied in analysis and planning of electric power grids, it is appropriate to discuss some items of which one should be aware. This concluding chapter will touch upon many such items that, though disparate, are intended to provide a broader perspective on matters concerning the reliability of electric power systems, and of the future of the field.

The main objective of this book is to address the subject of grid reliability. Structurally, generating plants supply power to the grid, and the distribution system transports power to the end users. One needs to be aware that there is considerable literature on topics such as collection and processing of outage or failure data for determination of generating unit availability [115]; methods for distribution system reliability [21, 116, 117]; methods and practices for reliability of industrial and commercial power systems [26]; and methods and practices for emergency and standby power supplies industrial and commercial systems [118]. Although the reliability principles and models described in this book are universally applicable, each segment of the electric industry has developed its own specific and sometimes unique methods and metrics.

Many professional organizations and regulatory entities have recommended and mandated industry practices, and continue to do so. The Institute of Electrical and Electronics Engineers (IEEE) has contributed to the development of the standards and recommended practices mentioned above. These standards and other documents have been developed collaboratively by experts from industry and academia and integrate mathematical principles and industry practices. These have been widely adopted by industry and by regulatory bodies. The cognizant regulatory body that regulates grid reliability and develops electric reliability standards on the North American continent is the North American Electric Reliability Corporation (NERC). NERC, originally founded in 1968 after the Northeast blackout of 1965, was appointed the Electric Reliability Organization (ERO) by the Federal Energy Regulatory Commission (FERC) in 2006, after the Northeast blackout of 2003, and continues

Electric Power Grid Reliability Evaluation: Models and Methods, First Edition. Chanan Singh, Panida Jirutitijaroen, and Joydeep Mitra.
© 2019 by The Institute of Electrical and Electronic Engineers, Inc. Published 2019 by John Wiley & Sons, Inc.

to develop standards for reliability and security of the electric infrastructure in North America[1]. NERC's responsibilities include investigation of blackout events so as to learn from these events and develop measures that may help prevent or mitigate future blackouts. The materials published by IEEE and NERC, in print and on their web sites, constitute a wealth of information on industry practices relating to reliability of electric power systems. NERC also maintains data collection systems and databases for resource availability, such as the Generating Availability Data System (GADS), the Transmission Availability Data System (TADS), and the Demand Response Availability Data System (DADS).

Before making conjectures about the future of the field and avenues for future research and development, it is worthwhile to discuss some of the factors that contributed to the recent resurgence of the importance of grid reliability and are shaping the future power systems. The late 1990s witnessed a worldwide movement toward restructuring the electric utility industry. In the US, the restructuring process and policies in many states produced several unintended consequences, most notably the following: (a) there was no clear allocation amongst market participants of operating practices that contributed to system reliability; (b) new operating practices, such as wheeling, caused unprecedented stresses on an already-aging infrastructure; and (c) system operators lacked experience with markets and new regulatory and corporate policies. Regulatory bodies and Independent System Operators have responded by developing new policies and products (ancillary services). There were many other factors that conspired to bring about the August 2003 blackout [119], but one of the most significant actions taken by NERC in response to this event was the formation of the Reliability Functional Model [120], a document that addressed the first of the three issues enumerated above by clearly and comprehensively defining "functional entities," encompassing standards developers, reliability service providers, and system planners and operators, and assigning specific functions to each functional entity to ensure system reliability. Penalties for noncompliance can range from large fines to suspension of an entity's ability to perform the function.

Two other movements have accompanied that of utility restructuring. One is the "clean energy" movement, which consists of replacing and augmenting conventional generation sources with "renewable resources," predominantly wind and solar. The other is the "smart grid" movement, which consists of the deployment of new communication, computing and control technologies to significantly enhance and augment the capabilities of traditional SCADA (Supervisory Control and Data Acquisition) systems. These three industry

---

1 When originally formed, this entity was called the "North American Electric Reliability Council," and most of its standards and recommendations were voluntarily adopted by utilities in the US, Canada, and Mexico. In 2007, following the appointment of NERC as the ERO in the US, the "Council" in the name was changed to "Corporation," and since 2006 all NERC standards have been enforceable in the US and have been adopted in many parts of Canada and Mexico.

trends have impacted grid reliability over the last two decades in unprecedented ways, and brought the issue to the forefront. Understanding these elements will assist the reader in anticipating emerging challenges in grid reliability and consequent directions in research on the field. Some of these are outlined below.

Reliability issues resulting from interaction of power delivery system with communication and control elements. Also referred to as reliability of cyber-physical systems, this is a complex and emergent topic. The interaction of physical and cyber systems introduce interdependencies that are typically difficult to handle in reliability models. Examples of the early work in this area are [121–124].

Reliability impacts of dynamic behavior of renewable resources. Most renewable resources have little or no inertia, and with increasing penetration and displacement of conventional generation, the system dynamics are affected, which in turn impact system reliability. References [125–129] report developments in this area.

Impact of emerging operating paradigms. Traditionally, operation of generating resources has been based on load following, i.e., dispatching generation to track system loads. However, with increasing proliferation of renewable and nondispatchable generation, and the increasing penetration of demand response and flexible loads, the industry appears to be moving in the direction of adopting a paradigm of generation following, where loads are controlled to track available generation. This would necessitate development of more sophisticated load models or "flexible resource" models. Customer participation is being increasingly enabled as two-way communication is integrated into advanced metering infrastructures. An example of early work in this area is [130].

Modeling and integration of storage technologies. Energy storage resources are gaining importance due to their ability to assist the penetration of renewable resources. With appropriate sizing and control, a storage device can assist a variable resource in becoming dispatchable, or, conversely, assist a load in being flexible. At the time of this writing, there is a strong thrust toward the development of utility-scale storage technologies. The emergence of viable technologies will necessitate the development of suitable models and integration methods. Early work in this area is reported in [44, 131–134].

Occasionally, there are breakthrough developments in the industry. More generally, however, we encounter continuous and incremental developments in response to changes in generating portfolios, new operating policies, etc. The following is an illustration of such incremental development. In the early days of wind generation, when total wind power production was low, the grid was able to accept all wind power produced, and it was simple to represent wind generation as negative load. With increasing penetration, and the implementation of pitch control in turbine blades, it became necessary and possible to "spill" wind production, and suitable models had to be developed (such as those described in Chapter 13).

Regardless of how the grid evolves, one thing is clear: grid reliability methods and models will be necessary to understand the impact of adopting new technologies, operating strategies and planning and operating decisions. We believe that in the future there will be increased complexities and interdependencies between the various segments of the grid. Some of the models described here will continue to be used, whereas others may need to be modified or enhanced to accommodate the changing scenarios. We have placed strong emphasis on the fundamentals so that the reader can acquire the capability to perform these modifications and enhancements. At least that is our hope!

Bibliography

1 Environmental Energy Technologies Division, Energy Analysis Department, Ernest Orlando Lawrence Berkeley National Laboratory, "Updated value of service reliability for electric utility customers in the United States," Ernest Orlando Lawrence Berkeley National Laboratory, Tech. Rep., January 2015.

2 C. Singh and A. D. Patton, "Concepts for calculating frequency of system failure," *IEEE Trans. Reliab.*, vol. 29, no. 4, pp. 336–338, 1980.

3 J. D. Hall, R. J. Ringlee, and A. J. Wood, "Frequency and duration methods for power system reliability calculations: I—generation system model," *IEEE Trans. Power App. Syst.*, vol. PAS-87, no. 9, pp. 1787–1796, 1968.

4 R. Billinton and C. Singh, "Generating capacity reliability evaluation in interconnected systems using a frequency and duration approach Part I. Mathematical analysis," *IEEE Trans. Power App. Syst.*, vol. PAS-90, no. 4, pp. 1646–1654, 1971.

5 R. Billinton and C. Singh, "System Load Representation in Generating Capacity Reliability Studies Part II. Applications and Extensions," *IEEE Trans. Power App. Syst.*, vol. PAS-91, no. 5, pp. 2133–2143, 1972.

6 A. K. Ayoub and A. D. Patton, "A frequency and duration method for generating system reliability evaluation," *IEEE Trans. Power App. Syst.*, vol. PAS-95, no. 6, pp. 1929–1933, 1976.

7 C. Singh, "Forced frequency-balancing technique for discrete capacity systems," *IEEE Trans. Reliab.*, vol. 32, no. 4, pp. 350–353, 1983.

8 A. C. G. Melo, M. V. F. Pereira, and A. M. L. da Silva, "A conditional probability approach to the calculation of frequency and duration indices in composite reliability evaluation," *IEEE Trans. Power Syst.*, vol. 8, no. 3, pp. 1118–1125, 1993.

9 J. Mitra and C. Singh, "Pruning and simulation for determination of frequency and duration indices of composite power systems," *IEEE Trans. Power Syst.*, vol. 14, no. 3, pp. 899–905, 1999.

Electric Power Grid Reliability Evaluation: Models and Methods, First Edition. Chanan Singh, Panida Jirutitijaroen, and Joydeep Mitra.
© 2019 by The Institute of Electrical and Electronic Engineers, Inc. Published 2019 by John Wiley & Sons, Inc.

10 A. D. Patton, C. Singh, and M. Sahinoglu, "Operating considerations in generation reliability modeling—an analytical approach," *IEEE Trans. Power App. Syst.*, vol. PAS-100, no. 5, pp. 2656–2663, 1981.

11 R. Billinton and C. L. Wee, "A frequency amd duration approach for interconnected system reliability evaluation," *IEEE Trans. Power App. Syst.*, vol. PAS-101, no. 5, pp. 1030–1039, 1982.

12 C. Singh and R. Billinton, *System Reliability, Modelling and Evaluation*. Hutchinson, 1977.

13 C. Singh, "On the behaviour of failure frequency bounds," *IEEE Trans. Reliab.*, vol. 26, no. 1, pp. 63–66, 1977.

14 J. C. Helton and F. J. Davis, "Latin hypercube sampling and the propagation of uncertainty in analyses of complex systems," *Reliab. Eng. Syst. Safety*, vol. 81, no. 1, pp. 23–67, 2003.

15 M. D. McKay, R. J. Beckman, and W. J. Conover, "A comparison of three methods for selecting values of input variables in the analysis of output from a computer code," *Technometrics*, vol. 21, no. 2, pp. 239–245, 1979.

16 D. P. Kroese, R. Y. Rubinstein, and P. W. Glynn, "The cross-entropy method for estimation," *Handbook of Statistics: Machine Learning: Theory and Applications*, North Holland, vol. 31, pp. 19–34, 2013.

17 A. M. L. da Silva, R. A. Fernandez, and C. Singh, "Generating capacity reliability evaluation based on Monte Carlo simulation and cross-entropy methods," *IEEE Trans. Power Syst.*, vol. 25, no. 1, pp. 129–137, 2010.

18 J. T. Linderoth, A. Shapiro, and S. J. Wright, "The empirical behavior of sampling methods for stochastic programming," *Ann. Oper. Res.*, vol. 142, no. 1, pp. 215–241, 2006.

19 B. Verweij, S. Ahmed, A. J. Kleywegt, G. Nemhauser, and A. Shapiro, "The sample average approximation method applied to stochastic routing problems: A computational study," *Comput. Optim. Appl.*, vol. 24, no. 2–3, pp. 289–333, February 2003.

20 M. Liefvendahl and R. Stocki, "A study on algorithms for optimization of Latin hypercubes," *J. Stat. Plan. Inference*, vol. 136, no. 9, pp. 3231–3247, 2006.

21 "IEEE guide for electric power distribution reliability indices," *IEEE Std 1366-2003 (Revision of IEEE Std 1366-1998)*, 2004.

22 System Protection and Controls Committee, "Reliability fundamentals of system protection," NERC, Tech. Rep., 2010.

23 System Protection and Controls Committee, "Protection system maintenance," NERC, Tech. Rep., 2007.

24 C. Singh and A. D. Patton, "Models and concepts for power system reliability evaluation including protection-system failures," *Int. J. Elec. Power Energy Sys.*, vol. 2, no. 4, pp. 161–168, 1980.

25 J. D. Grimes, "On determining the reliability of protective relay systems," *IEEE Trans. Reliab.*, vol. 19, no. 3, pp. 82–85, 1970.

26 "IEEE recommended practice for the design of reliable industrial and commercial power systems (IEEE Gold Book)," *IEEE Std 493-1997*, Aug 1998.

27 M. P. Bhavaraju, R. Billinton, G. L. Landgren, M. F. McCoy, and N. D. Reppen, "Proposed definitions of terms for reporting and analyzing outages of electrical transmission and distribution facilities and interruptions," *IEEE Trans. Power App. Syst.*, vol. PAS-87, no. 5, pp. 1318–1323, 1968.

28 J. Endrenyi, *Reliability Modeling in Electric Power Systems*. John Wiley & Sons, 1978.

29 N. S. Rau, C. Necsulescu, K. F. Schenk, and R. B. Misra, "Reliability of interconnected power systems with correlated demands," *IEEE Trans. Power App. Syst.*, vol. PAS-101, no. 9, pp. 3421–3430, 1982.

30 D. J. Levy and E. P. Kahn, "Accuracy of the Edgeworth approximation for LOLP calculations in small power systems," *IEEE Trans. Power App. Syst.*, vol. PAS-101, no. 4, pp. 986–996, 1982.

31 G. Gross, N. V. Garapic, and B. McNutt, "The mixture of normals approximation technique for equivalent load duration curves," *IEEE Trans. Power Syst.*, vol. 3, no. 2, pp. 368–374, 1988.

32 W. Tian, D. Sutanto, Y. Lee, and H. Outhred, "Cumulant based probabilistic power system simulation using Laguerre polynomials," *IEEE Trans. Energy Convers.*, vol. 4, no. 4, pp. 567–574, 1989.

33 P. Jorgensen, "A new method for performing probabilistic production simulations by means of moments and Legendre series," *IEEE Trans. Power Syst.*, vol. 6, no. 2, pp. 567–575, 1991.

34 C. Singh and J. Kim, "A continuous, distribution approach for production costing," *IEEE Trans. Power Syst.*, vol. 9, no. 3, pp. 1471–1477, 1994.

35 Reliability Test System Task Force of the Application of Probability Methods Subcommittee, "IEEE reliability test system," *IEEE Trans. Power App. Syst.*, vol. PAS-98, no. 6, pp. 2047–2054, 1979.

36 C. Singh, A. D. Patton, A. Lago-Gonzalez, A. R. Vojdani, G. Gross, F. F. Wu, and N. J. Balu, "Operating considerations in reliability evaluation of interconnected systems—an analytical approach," *IEEE Trans. Power Syst.*, vol. 3, pp. 123–129, 1988.

37 M. R. Anderberg, *Cluster Analysis for Applications: Probability and Mathematical Statistics*. Academic Press, 2014.

38 C. Singh and Q. Chen, "Generation system reliability evaluation using a cluster based load model," *IEEE Trans. Power Syst.*, vol. 4, no. 1, pp. 102–107, 1989.

39 A. Lago-Gonzalez and C. Singh, "The extended decomposition-simulation approach for multi-area reliability calculations," *IEEE Trans. Power Syst.*, vol. 5, no. 3, pp. 1024–1031, 1990.

40 W. F. Tinney and W. L. Powell, "The REI approach to power network equivalents," in *Proceedings of Power Industry Computer Applications Conference*, May 1977, pp. 314–320.

41 S. C. Savulescu, "Equivalents for security analysis of power systems," *IEEE Trans. Power App. Syst.*, vol. PAS-100, no. 5, pp. 2672–2682, 1981.

42 A. Patton and S. Sung, "A transmission network model for multi-area reliability studies," *IEEE Trans. Power Syst.*, vol. 8, no. 2, pp. 459–465, 1993.

43 J. Mitra and C. Singh, "Incorporating the DC load flow model in the decomposition-simulation method of multi-area reliability evaluation," *IEEE Trans. Power Syst.*, vol. 11, no. 3, pp. 1245–1254, 1996.

44 J. Mitra and M. R. Vallem, "Determination of storage required to meet reliability guarantees on island-capable microgrids with intermittent sources," *IEEE Trans. Power Syst.*, vol. 27, no. 4, pp. 2360–2367, 2012.

45 R. Billinton and W. Li, *Reliability Assessment of Electric Power Systems Using Monte Carlo Methods*. Springer, 2013.

46 R. Billinton and M. P. Bhavaraju, "Transmission planning using a reliability criterion, Part I: A reliability criterion," *IEEE Trans. Power App. Syst.*, vol. PAS-89, no. 1, pp. 28–34, 1970.

47 M. V. F. Pereira, L. M. V. G. Pinto, G. C. Oliveira, and S. H. F. Cunha, "Composite system reliability evaluation methods," *Final Report on Research Project 2473-10, EPRI EL-51*, no. 1, pp. 28–34, 1987.

48 T. A. Mikolinnas and B. F. Wollenberg, "An advanced contingency selection algorithm," *IEEE Trans. Power App. Syst.*, vol. PAS-100, no. 2, pp. 608–617, 1981.

49 C. Singh and G. Chintaluri, "Reliability evaluation of interconnected power systems using a multi-parameter gamma distribution," *Int. J. Elec. Power Energy Syst.*, vol. 17, no. 2, pp. 151–160, 1995.

50 J. Mitra and C. Singh, "Capacity assistance distributions for arbitrarily configured multi-area networks," *IEEE Trans. Power Syst.*, vol. 12, no. 4, pp. 1530–1535, 1997.

51 A. Meliopoulos, A. Bakirtzis, and R. Kovacs, "Power system reliability evaluation using stochastic load flows," *IEEE Trans. Power App. Syst.*, vol. PAS-103, no. 5, pp. 1084–1091, 1984.

52 T. Karakatsanis and N. Hatziargyriou, "Probabilistic constrained load flow based on sensitivity analysis," *IEEE Trans. Power Syst.*, vol. 9, no. 4, pp. 1853–1860, 1994.

53 K. Moslehi and F. Wu, "A method for bulk power system reliability evaluation based on local coherency," *Elec. Power Syst. Res.*, vol. 7, no. 4, pp. 307–319, 1984.

54 P. Doulliez and E. Jamoulle, "Transportation networks with random arc capacities," *Rev. Fr. Inform. Rech. O*, vol. 6, no. 3, pp. 45–59, 1972.

55 D. P. Clancy, G. Gross, and F. F. Wu, "Probabilitic flows for reliability evaluation of multiarea power system interconnections," *Int. J. Elec. Power Energy Syst.*, vol. 5, no. 2, pp. 101–114, 1983.

56 C. Singh and Z. Deng, "A new algorithm for multi-area reliability evaluation simultaneous decomposition-simulation approach," *Elec. Power Syst. Res.,* vol. 21, no. 2, pp. 129–136, 1991.

57 C. Singh and J. Mitra, "Composite system reliability evaluation using state space pruning," *IEEE Trans. Power Syst.,* vol. 12, no. 1, pp. 471–479, 1997.

58 K. Clements, B. Lam, D. Lawrence, T. Mikolinnas, N. Reppen, R. Ringlee, and B. Wollenberg, "Transmission-system reliability methods. Volume 1. Mathematical models, computing methods, and results. Final report," Power Technologies, Inc., Tech. Rep., July 1982.

59 C. Singh and J. Mitra, "Reliability analysis of emergency and standby power systems," *IEEE Ind. Appl. Mag.,* vol. 3, no. 5, pp. 41–47, 1997.

60 X. Luo, C. Singh, and A. D. Patton, "Power system reliability evaluation using learning vector quantization and Monte Carlo simulation," *Elec. Power Syst. Res.,* vol. 66, no. 2, pp. 163–169, Aug. 2003.

61 G. J. Anders, *Probability Concepts in Electric Power Systems.* John Wiley & Sons, 1990.

62 S. Huang, "Effectiveness of optimum stratified sampling and estimation in Monte Carlo production simulation," *IEEE Trans. Power Syst.,* vol. 12, no. 2, pp. 566–572, 1997.

63 M. Mazumdar, "Importance sampling in reliability estimation," in *Conference on Reliability and Fault Tree Analysis,* Berkeley, California, 3 Sep 1974, pp. 153–163; 1975.

64 R. A. Gonzalez-Fernndez, A. M. L. da Silva, L. C. Resende, and M. T. Schilling, "Composite systems reliability evaluation based on Monte Carlo simulation and cross-entropy methods," *IEEE Trans. Power Syst.,* vol. 28, no. 4, pp. 4598–4606, 2013.

65 J. Mitra, "Models for reliability evaluation of multi-area and composite systems," Ph.D. dissertation, Texas A&M University, 1997.

66 A. C. G. Melo, M. V. F. Pereira, and A. M. L. da Silva, "Frequency and duration calculations in composite generation and transmission reliability evaluation," *IEEE Trans. Power Syst.,* vol. 7, no. 2, pp. 469–476, 1992.

67 L. Wang and C. Singh, "Population-based intelligent search in reliability evaluation of generation systems with wind power penetration," *IEEE Trans. Power Syst.,* vol. 23, no. 3, pp. 1336–1345, 2008.

68 N. Samaan and C. Singh, "Adequacy assessment of power system generation using a modified simple genetic algorithm," *IEEE Trans. Power Syst.,* vol. 17, no. 4, pp. 974–981, 2002.

69 N. Samaan and C. Singh, "Assessment of the annual frequency and duration indices in composite system reliability using genetic algorithms," in *Power Engineering Society General Meeting, 2003, IEEE,* vol. 2, 2003, pp. 692–697.

70 J. Kennedy, "Particle swarm optimization," in *Encyclopedia of Machine Learning,* 2011, pp. 760–766.

71 S. B. Patra, J. Mitra, and R. Earla, "A new intelligent search method for composite system reliability analysis," in *IEEE PES Transmission and Distribution Conference and Exhibition*, 2006, pp. 803–807.

72 M. T. Schilling, J. C. S. Souza, A. P. A. da Silva, and M. B. Do Coutto Filho, "Power systems reliability evaluation using neural networks," *Int. J. Eng. Intell. Syst.*, vol. 4, no. 4, pp. 219–226, 2001.

73 C. Singh, X. Luo, and H. Kim, "Power system adequacy and security calculations using Monte Carlo simulation incorporating intelligent system methodology," in *Proceedings of International Conference on Probabilistic Methods Applied to Power Systems*, 2006, pp. 1–9.

74 X. Luo, C. Singh, and A. Patton, "Power system reliability evaluation using self-organizing map," in *Power Engineering Society Winter Meeting, 2000, IEEE*, vol. 2, 2000, pp. 1103–1108.

75 P. Yuanidis, M. A. Styblinski, D. R. Smith, and C. Singh, "Reliability modeling of flexible manufacturing systems," *Microelectron. Reliab.*, vol. 34, no. 7, pp. 1203–1220, 1994.

76 M. Benidris and J. Mitra, "Reliability and sensitivity analysis of composite power systems under emission constraints," *IEEE Trans. Power Syst.*, vol. 29, no. 1, pp. 404–412, 2014.

77 J. Zhu and M. Chow, "A review of emerging techniques on generation expansion planning," *IEEE Trans. Power Syst.*, vol. 12, no. 4, pp. 1722–1728, 1997.

78 G. Infanger, *Planning Under Uncertainty: Solving Large-Scale Stochastic Linear Programs*. Boyd & Fraser, 1993.

79 P. Jirutitijaroen and C. Singh, "Reliability constrained multi-area adequacy planning using stochastic programming with sample-average approximations," *IEEE Trans. Power Syst.*, vol. 23, no. 2, pp. 504–513, 2008.

80 P. Jirutitijaroen and C. Singh, "Composite-system generation adequacy planning using stochastic programming with sample-average approximation," *Proc. of 16th Power Syst. Comput. Conf.*, Glasgow, Scotland, UK, July 14–18, 2008.

81 P. Jirutitijaroen and C. Singh, "Unit availability considerations in composite-system generation planning," *Proc. of 10th Int. Conf. on Prob. Method Appl. to Power Syst.*, Glasgow, Scotland, UK, July 14–18, 2008.

82 J. R. Birge and F. Louveaux, *Introduction to Stochastic Programming*. Duxbury Press, 1997.

83 J. L. Higle and S. Sen, *Stochastic Decomposition: A Statistical Method for Large Scale Stochastic Linear Programming*. Kluwer Academic Publishers, 1996.

84 P. Jirutitijaroen and C. Singh, "Comparative study of system-wide reliability-constrained generation expansion problem," *Proc. of 3rd Int. Conference on Electric Utility Deregulation and Restructuring and Power Technologies*, 2008.

85 L. Lawton, M. Sullivan, K. V. Liere, A. Katz, and J. Eto, "A framework and review of customer outage costs: Integration and analysis of electric utility outage cost surveys," *Lawrence Berkeley National Laboratory. Paper LBNL-54365.*, 2003.

86 W. K. Mak, D. P. Morton, and R. K. Wood, "Monte Carlo bounding techniques for determining solution quality in stochastic programs," *Oper. Res. Lett.*, vol. 24, no. 1-2, pp. 47–56, 1999.

87 P. Tavner, C. Edwards, A. Brinkman, and F. Spinato, "Influence of wind speed on wind turbine reliability," *Wind Eng.*, vol. 30, no. 1, pp. 55–72, 2006.

88 J. Ge, M. Du, and C. Zhang, "A Study on Correlation of Wind Farms Output in the Large-Scale Wind Power Base," in *Proc. of the 4th International Conference on Electric Utility Deregulation and Restructuring and Power Technologies*, Weihai, Shandong, 2011, pp. 1316–1319.

89 B. Hasche, "General statistics of geographically dispersed wind power," *Wind Energy*, vol. 13, no. 8, pp. 773–784, Nov. 2010.

90 B. Ernst, Y.-H. Wan, and B. Kirby, "Short-Term Power Fluctuation of Wind Turbines: Analyzing Data from the German 250-MW Measurement Program from the Ancillary Services Viewpoint," in *Proc. of the Wind Power '99 Conference*, Burlington, Vermont, June 20–23, 1999, pp. 1–10.

91 Y. Wan, M. Milligan, and B. Parsons, "Output power correlation between adjacent wind power plants," *J. Solar Energy Eng.*, vol. 125, no. 4, pp. 551–555, 2003.

92 H. Louie, "Correlation and statistical characteristics of aggregate wind power in large transcontinental systems," *Wind Energy*, vol. 17, no. 6, pp. 793–810, 2014.

93 P. Kádár, "Evaluation of correlation the wind speed measurements and wind turbine characteristics," in *Proc. of the 8th International Symposium of Hungarian Researchers on Computational Intelligence and Informatics*, Budapest, Hungary, Nov. 15–17, 2007, pp. 15–17.

94 J. R. Ubeda and M. A. R. Rodriguez Garcia, "Reliability and production assessment of wind energy production connected to the electric network supply," *IEE Proc. Generat. Transm. Distrib.*, vol. 146, no. 2, pp. 169–175, 1999.

95 H. Sipeng, Z. Yangfei, L. Xianyun, and Y. Yue, "Equivalent wind speed model in wind farm dynamic analysis," in *4th International Conference on Electric Utility Deregulation and Restructuring and Power Technologies*, Weihai, Shandong, July 6–9, 2011, pp. 1751–1755.

96 S. Tanneeru, "Reliability Modeling of DG Clusters," Master's Thesis, New Mexico State University, 2008.

97 T. Nanahara, M. Asari, T. Maejima, T. Sato, K. Yamaguchi, and M. Shibata, "Smoothing effects of distributed wind turbines. Part 2. Coherence among power output of distant wind turbines," *Wind Energy*, vol. 7, no. 2, pp. 75–85, 2004.

98 S. Venkatraman *et al.*, "Integration of Renewable Resources," California ISO, Folsom, CA, Tech. Rep., August 2010.

99 K. Coughlin and J. H. K. Eto, "Analysis of wind power and load data at multiple time scales," Technical Report LBNL-4147E, U.S. Department of Energy, Tech. Rep., 2010.

100 H. Kim, C. Singh, and A. Sprintson, "Simulation and estimation of reliability in a wind farm considering the wake effect," *IEEE Trans. Sustain. Energy*, vol. 3, no. 2, pp. 274–282, 2012.

101 W. Wangdee and R. Billinton, "Considering load-carrying capability and wind speed correlation of WECS in generation adequacy assessment," *IEEE Trans. Energy Convers.*, vol. 21, no. 3, pp. 734–741, 2006.

102 B. Martin and J. Carlin, "Wind-load correlation and estimates of the capacity credit of wind power: An empirical investigation," *Wind Eng.*, vol. 7, no. 2, p. 79, 1983.

103 R. Perez, R. Seals, and R. Stewart, "Matching utility peak loads with photovoltaics," in *RENEW94 Conference, Stamford, CT, Northeast Sustainable Solar Energy Association*, Greenfield, MA, April 11–13, 1994.

104 S. Sulaeman, M. Benidris, J. Mitra, and C. Singh, "A wind farm reliability model considering both wind variability and turbine forced outages," *IEEE Trans. Sustain. Energy*, vol. 8, no. 2, pp. 629–637, 2017.

105 C. Singh and A. Lago-Gonzalez, "Reliability modeling of generation systems including unconventional energy sources," *IEEE Trans. Power App. Syst.*, vol. PAS-104, no. 5, pp. 1049–1056, 1985.

106 C. Singh and A. Bagchi, "Reliability analysis of power systems incorporating renewable energy sources," in *16th National Power Systems Conference*, Hyderabad, India, December 15th–17th, 2010.

107 S. Fockens, A. J. M. Van Wijk, W. C. Turkenburg, and C. Singh, "Reliability analysis of generating systems including intermittent sources," *Int. J. Elec. Power Energy Syst.*, vol. 14, no. 1, pp. 2–8, 1992.

108 A. D. Patton, A. K. Ayoub, C. Singh, G. L. Hogg, and J. W. Foster, "Modeling of unit operating considerations in generating-capacity reliability evaluation. Volume 1. Mathematical models, computing methods, and results." Electric Power Research Institute Report EPRI EL-2519, Vol 1, July 1982.

109 C. Singh and Y. Kim, "An efficient technique for reliability analysis of power systems including time dependent sources," *IEEE Trans. Power Syst.*, vol. 3, no. 3, pp. 1090–1096, 1988.

110 SAS Institute, "SAS user's guide: statistics," *5th Edition. SAS Institue Inc.*, Cary, NC, 1985.

111 W. Sarle, "SAS technical report A-108," *The Cubic Clustering Criterion. Cary, NC: SAS Institute*, 1983.

112 H. Kim and C. Singh, "Three dimensional clustering in wind farms with storage for reliability analysis," in *PowerTech (POWERTECH), 2013 IEEE Grenoble*. IEEE, Grenoble, France, Jun 16–20, 2013, pp. 1–6.

113 S. Fockens, A. J. M. Van Wijk, W. C. Turkenburg, and C. Singh, "A concise method for calculating expected unserved energy in generating system reliability analysis," *IEEE Trans. Power Syst.*, vol. 6, no. 3, pp. 1085–1091, 1991.

114 M. Ramezani, C. Singh, and M. R. Haghifam, "Role of clustering in the probabilistic evaluation of ttc in power systems including wind power generation," *IEEE Trans. Power Syst.*, vol. 24, no. 2, pp. 849–858, 2009.

115 "IEEE standard definitions for use in reporting electric generating unit reliability, availability, and productivity," *IEEE Std. 762-2006 (Revision of IEEE Std 762-1987)*, 2007.

116 A. A. Chowdhury and D. Koval, *Power Distribution System Reliability: Practical Methods and Applications*. Vol. 48. John Wiley & Sons, 2011.

117 R. E. Brown, *Electric Power Distribution Reliability*, 2nd ed. CRC Press, 2009.

118 "IEEE recommended practice for power system analysis (IEEE Brown Book)," *ANSI/IEEE Std 399-1980*, 1980.

119 US-Canada Power System Outage Task Force, *Final Report on the August 14, 2003 Blackout in the United States and Canada: Causes and Recommendations*. US-Canada Power System Outage Task Force, 2004.

120 Functional Model Working Group, *Reliability Functional Model Technical Document—Version 5*. NERC, Dec 2009.

121 B. Falahati, Y. Fu, and L. Wu, "Reliability assessment of smart grid considering direct cyber-power interdependencies," *IEEE Trans. Smart Grid*, vol. 3, no. 3, pp. 1515–1524, 2012.

122 H. Lei, C. Singh, and A. Sprintson, "Reliability modeling and analysis of IEC 61850 based substation protection systems," *IEEE Trans. Smart Grid*, vol. 5, no. 5, pp. 2194–2202, 2014.

123 H. Lei and C. Singh, "Power system reliability evaluation considering cyber-malfunctions in substations," *Elec. Power Syst. Res.*, vol. 129, pp. 160–169, 2015.

124 H. Lei and C. Singh, "Non-sequential monte carlo simulation for cyber-induced dependent failures in composite power system reliability evaluation," *IEEE Trans. Power Syst.*, vol. 32, no. 2, pp. 1064–1072, 2017.

125 D. Gautam, V. Vittal, and T. Harbour, "Impact of increased penetration of dfig-based wind turbine generators on transient and small signal stability of power systems," *Trans. Power Syst.*, vol. 24, no. 3, pp. 1426–1434, 2009.

126 E. Vittal, M. O'Malley, and A. Keane, "A steady-state voltage stability analysis of power systems with high penetrations of wind," *IEEE Trans. Power Syst.*, vol. 25, no. 1, pp. 433–442, 2010.

127 S. Eftekharnejad, V. Vittal, G. T. Heydt, B. Keel, and J. Loehr, "Impact of increased penetration of photovoltaic generation on power systems," *IEEE Trans. Power Syst.*, vol. 28, no. 2, pp. 893–901, 2013.

128 N. Nguyen and J. Mitra, "An analysis of the effects and dependency of wind power penetration on system frequency regulation," *IEEE Trans. Sustain. Energy*, vol. 7, no. 1, pp. 354–363, 2016.

129 N. Nguyen and J. Mitra, "Reliability of power system with high wind penetration under frequency stability constraint," *IEEE Trans. Power Syst.*, vol. 33, no. 1, pp. 985–994, 2018.

130 M. S. Modarresi, L. Xie, and C. Singh, "Reserves from controllable swimming pool pumps: Reliability assessment and operational planning," in *51st Hawaii International Conference on System Sciences (HICSS)*, Waikoloa Village, HI, January 3–6, 2018, pp. 1–10.

131 S. Sulaeman, Y. Tian, M. Benidris, and J. Mitra, "Quantification of storage necessary to firm up wind generation," *IEEE Trans. Ind. Appl.*, vol. 53, no. 4, pp. 3228–3236, 2017.

132 P. Xiong and C. Singh, "Optimal planning of storage in power systems integrated with wind power generation," *IEEE Trans. Sustain. Energy*, vol. 7, no. 1, pp. 232–240, 2016.

133 F. Alismail, P. Xiong, and C. Singh, "Optimal wind farm allocation in multi-area power systems using distributionally robust optimization approach," *IEEE Trans. Power Syst.*, vol. 33, no. 1, pp. 536–544, 2018.

134 Y. Xu and C. Singh, "Power system reliability impact of energy storage integration with intelligent operation strategy," *IEEE Trans. Smart Grid*, vol. 5, no. 2, pp. 1129–1137, 2014.

135 C. Singh and N. V. Gubbala, "An alternative approach to rounding off generation models in power system reliability evaluation," *Elec. Power Syst. Res.*, vol. 36, no. 1, pp. 37–44, 1996.

136 J. Mitra, "Reliability-based sizing of backup storage," *IEEE Trans. Power Syst.*, vol. 25, no. 2, pp. 1198–1199, 2010.

137 J. Mitra and C. Singh, "A hybrid approach to addressing the problem of noncoherency in multi-area reliability models," in *Proc. Power System Computation Conference*, Dresden, Aug. 20–23, 1996, pp. 1011–1017.

138 R. Ramakumar, *Engineering Reliability: Fundamentals and Applications*. Prentice Hall, 1996.

139 R. Billinton and R. N. Allan, *Reliability Evaluation of Engineering Systems: Concepts and Techniques*. Springer, 1983.

140 R. Billinton and R. N. Allan, *Reliability Evaluation of Power Systems*, 2nd ed. Springer, 1996.

141 R. Billinton and R. N. Allan, *Reliability Assessment of Large Electric Power Systems*. Kluwer Academic Publishers, 1988.

142 J. F. Manwell, J. G. McGowan, and A. L. Rogers, *Wind Energy Explained: Theory, Design and Application*. John Wiley & Sons, 2010.

143 D. Elmakias, *New Computational Methods in Power System Reliability*. Springer, 2008.

144 A. Patton, A. Ayoub, C. Singh, G. Hogg, and J. Foster, Vol 2: Computer Program Documentation, EPRI EL- 2519, Vol 2, July 1982.

145 B. S. Dhillon and C. Singh, *Engineering Reliability: New Techniques and Applications*, John Wiley & Sons Ltd. 1981.

146 M. T. Schilling and A. M. L. da Silva, "Conceptual investigation on probabilistic adequacy protocols: Brazilian experience," *IEEE Trans. Power Syst.*, vol. 29, no. 3, pp. 1270–1278, 2014.

147 M. T. Schilling, J. C. S. de Souza, and M. B. Do Coutto Filho, "Power system probabilistic reliability assessment: Current procedures in Brazil," *IEEE Trans. Power Syst.*, vol. 23, no. 3, pp. 868–876, 2008.

148 X. Liang, H. E. Mazin, and S. E. Reza, "Probabilistic generation and transmission planning with renewable energy integration," in *2017 IEEE/IAS 53rd Industrial and Commercial Power Systems Technical Conference (I&CPS)*, Niagara FAlls, ON Canada, May 7–11, 2017, pp. 1–9.

149 R. N. F. Filho, M. T. Schilling, J. C. O. Mello, and J. L. R. Pereira, "Topological reduction considering uncertainties," *IEEE Trans. Power Syst.*, vol. 10, no. 2, pp. 739–744, 1995.

150 M. T. Schilling, J. C. G. Praca, J. F. de Queiroz, C. Singh, and H. Ascher, "Detection of ageing in the reliability analysis of thermal generators," *IEEE Trans. Power Syst.*, vol. 3, no. 2, pp. 490–499, 1988.

151 M. Benidris, J. Mitra, and C. Singh, "Integrated evaluation of reliability and stability of power systems," *IEEE Trans. Power Syst.*, vol. 32, no. 5, pp. 4131–4139, 2017.

152 Y. Zhang, A. A. Chowdhury, and D. O. Koval, "Probabilistic wind energy modeling in electric generation system reliability assessment," *IEEE Trans. Ind. Appl.*, vol. 47, no. 3, pp. 1507–1514, 2011.

Index

Electric Power Grid Reliability Evaluation: Models and Methods, First Edition. Chanan Singh,
Panida Jirutitijaroen, and Joydeep Mitra.
© 2019 by The Institute of Electrical and Electronic Engineers, Inc. Published 2019 by John Wiley & Sons, Inc.